WHEN THE PLANET RAGES

WHEN THE PLANET RAGES

Natural Disasters, Global Warming, and the Future of the Earth

Charles Officer
Jake Page

OXFORD
UNIVERSITY PRESS

2009

OXFORD
UNIVERSITY PRESS

Oxford University Press, Inc., publishes works that further
Oxford University's objective of excellence
in research, scholarship, and education.

Oxford New York
Auckland Cape Town Dar es Salaam Hong Kong Karachi
Kuala Lumpur Madrid Melbourne Mexico City Nairobi
New Delhi Shanghai Taipei Toronto

With offices in
Argentina Austria Brazil Chile Czech Republic France Greece
Guatemala Hungary Italy Japan Poland Portugal Singapore
South Korea Switzerland Thailand Turkey Ukraine Vietnam

First published as *Tales of the Earth* by Oxford University Press, Inc., 1993.
First issued as an Oxford University Press paperback, 1994.
198 Madison Avenue, New York, NY 10016

www.oup.com

Library of Congress Cataloging-in-Publication Data
Officer, Charles B.
When the planet rages : natural disasters, global warming,
and the future of the earth / Charles Officer and Jake Page.
p. cm.
Rev. ed. of : Tales of the earth. New York : Oxford University Press, c1993.
Includes bibliographical references and index.
ISBN: 978-0-19-537701-9 (pbk.)
1. Earth. 2. Environmental degradation.
[1. Nature–Effect of human beings on.]
I. Page, Jake. II. Officer, Charles B. Tales of the earth. III. Title.
QB631.O34 2009
550—dc22 2009014874

1 3 5 7 9 8 6 4 2
Printed in the United States of America
on acid-free paper

Preface to the Revised Edition

In 1993, when the manuscript for this book went off to Oxford University Press to be published as *Tales of the Earth*, the town of Chelsea, Iowa, some seventy miles north of Des Moines, was largely under water. The state and the region had suffered what appeared to be a 100-year flood. The Iowa River was profoundly swollen, and the spring itself had been unusually rainy. In all, throughout the region, fifty people died and fifty thousand homes were destroyed in nine Midwestern states.

In the aftermath, the federal government offered townspeople throughout the Mississippi drainage whose houses and livelihoods had suffered from flooding the funds needed to relocate themselves to higher ground. This generosity was sparked by the idea that the federal government should not be asked to pay for flood damage year after year in one part of the drainage or another, such flooding being essentially inevitable.

For the people of Chelsea who had grown up and spent their entire lives there, this was a painful decision. In the end, only 40 percent of the townspeople moved. Fifteen years later, as this revised edition was being prepared, Chelsea was again under some six feet of water. As one resident told the *New York Times*, exaggerating some, "I'm not sure how much wisdom there is in staying because these are floods that are supposed to come every 500 years and they're coming every 15 years."

It was in fact an altogether terrible year for Iowans among others in the American Midwest. Tornados of unusual lethality had repeatedly swept through the state, in one instance demolishing a Boy Scout camp. Aerial photographs showed towns across the state that looked like piles of splinters strewn angrily across the land. Indeed, the National Oceanic and Atmospheric Administration's Storm Prediction Center reported that by June 2008, tornados had taken 115 lives in the United States, the most since 1998. It was the third super-outbreak of tornados since 1974. Major storms were occurring in the Midwest every three or four days, many of them spawning tornados.

Back in Iowa, the wet weather and the floods had done long-term damage to croplands there, what with the equivalent of the usual *annual* rainfall occurring in one day in some places, washing away not just topsoil but also nitrogen fertilizers that make their way ever southward via the great and grand Mississippi drainage into the Gulf of Mexico, expanding the oxygen-free dead zone that exists there thanks to human agricultural activity. We live in a world of unintended consequences and occasional retaliation by the planet.

Flooding in the American Midwest in 2008 was perhaps the least lethal environmental news one might read in the papers or on line or see on television. The spring of 2008 also brought into our homes news of the tremendously widespread damage of the Sichuan earthquake in China, infamous for causing countless school buildings—built to a lesser code than most urban and suburban homes, it turned out—to collapse upon their students.

It was also the time when a cyclone named Nargis (meaning quite awfully in Urdu "daffodil") slammed into the Irriwaddy wetlands of southern Myanmar on May 2 with winds of up to 150 miles per hour and raged up through the main rice-growing lands of the nation formerly called Burma, leaving tens of thousand homeless, destitute, and desperate for the aid of sympathetic nations—which was freely offered but largely forbidden by the ruling junta, whose officers and soldiers ran off with whatever aid did get there. Some humans are also natural disasters. The final tally of human life lost and property ruined may never be known.

One had the sense in 2008 that the world had gone a bit mad, the Earth contorting itself in multiple seizures, and the Earth's denizens were in some cases doing a relatively lousy job of handling it. That spring, the weather channel had to all practical purposes become the natural disaster channel. And not just all of a sudden or all that recently. Few Americans would forget the video of the waters brought on by Hurricane Katrina rising in New Orleans until they breached the levees and water, unwanted and soon disgusting, filled a large part of the city. Nor will anyone who saw any of this on television forget the pictures of the people waiting to be plucked from their roofs by late-arriving rescue workers, or crowding into two evacuation centers already crammed with far too many people in degrading circumstances closest to those politely called feedlots. The slow federal response to this catastrophe may well be taken by future historians as the seed of downfall of the George W. Bush administration.

Thus does nature bring about its effects on history. Nature is especially noisy from time to time, and it is such particularly noisy and scary events that are the subject of this book. In the interval since its first publication we have, as noted, experienced some devastating events, sent to harass us by what some prefer to think of as the gentle, magical Mother Earth. Certainly this planet can be thought of that way without too much harm, so long as we not only honor it with acts of stewardship but also realize that Mother Earth does experience what seem to be unprovoked rages, like the fury of some ancient Egyptian goddess. It is the only Earth we have, or will have long into the future, and for all its comfort, for all its wonder, it can be a terribly dangerous place to live.

And recently it has been established beyond any doubt aside from that of lunatics that we humans have been abusing the Earth to the point where it seems to have entered a grumpy period of retaliation. It has now been shown beyond a doubt that human endeavors can actually change the climate of this huge place called planet Earth. It may seem impossible that something as puny as humans compared to the breadth and heft of the earth can have so wide-reaching an effect. But we dwell in an extremely thin skin of the earth, called the biosphere. Even before most of us got our heads into the problem formerly called global warming (it is now global climate change because its effects will be far from uniform), we realized that chlorofluorocarbons that we let escape from air conditioners, refrigerators, and other devices were making a huge hole in the ozone layer over Antarctica and a thinning of the ozone layer—globally a huge anthropogenic, or human-caused, effect.

We all soon learned that the ozone layer protects all of life from the sun's destructive radiation called ultraviolet, causing, among other things, skin cancer as the punishment for getting a lovely tan. We also, as a world population, realized that there was something we could do about that, and we did it, raising hell with our governments and soon getting rid of chlorofluorocarbons on a global basis. Obligingly, the ozone layer is growing again, healing itself. By the latter part of this century, it should be back to normal, and this is a cause for celebration—or at least two cheers. Two because no one modeling our climate has yet come up with an idea of exactly how a complete normal ozone layer will affect global climate change. It is yet to be included in the models climatologists build to forecast such things. There is a possibility that it might be helpful, at least in the southern hemisphere, and it certainly would be helpful to know that.

Mother Earth, it turns out, is a very complicated personality. Often and always somewhere, she is benign. Even nurturing. But she suffers great tectonic and other forces, what we could imagine perhaps as pain. And we can add to her problems. Tug on her skirt the wrong way, and it can tatter her overall attire in unpredictable ways. Poke her and she can explode, as these tales of the earth—some ancient and some new—we hope make clear. Maybe this book has already and will continue even to help impel us to make more lasting accommodations to this astonishing and singular Earth—and its often violent mood swings.

Charles Officer and Jake Page
Fall 2008

Preface to the First Edition

People who try to look at the whole Earth are proliferating: climatologists, macroecologists, international statesmen, multinational corporate officers... the list grows. Among the first to do this were geologists, who tend to operate in a unique realm of time.

The far-sighted CEO may be looking at the world from the standpoint of five-year, maybe twenty-year, investments. Statesmen seek peace in our time, but "our time" often turns out to be a decade or two. Cosmologists, on the other hand, look at such vast stretches of time and space that an era on Earth is a barely noticeable blip. Quantum mechanics deals in unimaginable fractions of nanoseconds.

Geologists are located somewhere between the cosmologist and the CEO. Theirs is a leisurely time frame in which continents have ample time to drift around on the face of the globe and mountains to arise where once there was an inland sea, only to be eroded down to a level plain. In this kind of broad time sense, what appear to be loud, massive, sudden, and utterly catastrophic events don't really startle a geologist (as scientist) very much. Such interesting, even telling, events punctuate the long history of Earth. And given that so many epochal events have occurred in that history, the geologist has nothing but the utmost respect for the power of the Earth to in a sense manage its own affairs regardless of human desires and dreams.

Yet mankind has suddenly become ubiquitous and dominant over all the Earth's flora and fauna. Humans now control the world as a habitat for life, and this has come about in a blink of geologic time. And we are regularly bombarded with the information that we are not doing a very good job of managing this platform for life. Much of the information is conflicting: apocalypse now, or maybe never. We are jostled between clarion calls to panic and suggestions that we simply forget the whole thing. Why the conflicting advice? In part, because we stand in the middle of a series of global experiments undertaken on a time scale never before attained. It is hard to monitor an experiment from inside the test tube. We make mathematical models in hopes of seeing a solution, but how can you model something if you don't know how it works? We have not yet spent enough time finding out how the Earth works. When an environmental problem arises, we want answers now, not after a two-decade research program. We need to embark on a new scientific voyage based on patience and long-term commitment.

It is comforting to think of the Earth as a stable place, *terra firma*, one that we can count on and that won't cause us undue stress. But of course we know that's not the case—and on a variety of time scales. We're aware of yearly climatic changes leading to bumper crops or droughts, of floods, of volcanic eruptions and earthquakes. We have read about the Ice Ages, about vast extinction events when the likes of dinosaurs vanished from this plane of existence. And we hear ominous things about how our own activities are now changing the nature of the Earth.

Even in a time of unprecedented change—that series of experiments—there is validity to the adage that "the key to the future is the past." In this book, we look at some of the great events in geologic history, many of them occurring before our arrival, but many of them instances of nature making trouble for humanity, and still others of humanity making trouble for nature. These "tales of the Earth" are related here in an attempt to explain what caused them (insofar as scientists understand the causes) and to put our own current travails in perspective. Part of the process of telling these tales is to look at how scientists and other individuals have gone about the business of understanding such events. Not a few of these, people like Benjamin Franklin, Thomas Jefferson, Voltaire, Rousseau, and Plato, might be called Renaissance men. Such men were interested in mankind, in nature and in all their ramifications—oddballs from the standpoint of today's tendency toward specialized knowledge. A good deal of the intent behind this book (as with any book of popular science) is to bring nonspecialists into the picture, in a sense to put people in charge. Our present environmental concerns are too important to be left to the scientists alone.

Hanover, N.H. C. O.
Corrales, N.M. J. P.
August 1992

Acknowledgments

Much of this book contains stories and anecdotes of the effects of both Nature and Man on our environment. In preparing the book we have relied, in part, on previous works by others, and we refer the reader to the References for further and more detailed information on these various tales.

We have also used descriptions and observations by the following individuals directly involved with particular events: The commentary by Thomas Raffles and Samuel Mitchell on the Tambora eruption came from the book by H. and E. Stommel on volcano weather and that by the Moravian landowner from the book by J. D. Post on the last great subsistence crisis in Europe. The commentary by Benjamin Franklin on the environmental effects of the Laki eruption came from the article by H. Sigurdsson on the same subject. The commentary by the survivor of the Krakatoa eruption came from the book by T. Simkin and R. S. Fiske on Krakatoa. Plato's account of the end of Atlantis came from the book by J. V. Luce on the same subject. The anecdote on Enrico Caruso's reaction to the San Francisco earthquake came from an article by E. Sorel. The eyewitness accounts of the New Madrid earthquake came from the book by M. L. Fuller. The commentaries on the Lisbon earthquake came from the book by T. D. Kendrick and the excerpt from Voltaire's poem on the earthquake from the translation by A. Hecht. The translation of the Gilgamesh legend of the Great Flood is from the book by D. Rosenberg on world mythology. The eyewitness description of the reappearance of Halley's comet in 1910 came from the book by R. Etter and

S. Scheider on the same subject. The reports on the effects of the Little Ice Age are from the books by J. M. Grove and H. H. Lamb. The commentaries on the early findings of the American mastodon and woolly mammoth remains are from the book by J. C. Greene. The descriptions by John Evelyn, Arthur Conan Doyle, and others on London smog are from the book by W. Wise on the killer smog. The specific wording of Thomas Malthus's dismal theorem came from the book by D. Winch on Malthus.

We have also benefited from conversations, comments, suggestions, and other forms of assistance from a number of individuals. These include Kenneth Belitz, Robert Berner, Wendy Berryman, Peter Bien, Jere Daniell, Norman Doenges, Charles Drake, Sandra Fullington, Anthony Hallam, Sheila Harvey, Daniel Lynch, John Lyons, Scott Mahler, Thomas McFarland, Grace Morse, William Oman, Naomi Oreskes, Eric Posmentier, Robert Reynolds, and Leslie Sonder. We should also like to acknowledge the assistance from the librarians at Dartmouth College, particularly those at the Kresge Physical Sciences Library; the drafting and photographic personnel at the Dartmouth Medical School; our editor, Joyce Berry, and agent, Robert Ducas; and most particularly our families.

Contents

Part Two Changes in Climate and Life on Earth

Part Three Our Effect On Nature

WHEN THE PLANET RAGES

PART ONE

NATURE'S EFFECT ON MAN

1

The Earth Is Still Hot and Mobile

People in Maryland knew something was up when the snows of late spring were brown, blue, even red. It was 1816, an unusually cold year. Brown snow fell in Hungary that year, and for the citizens of Taranto in southern Italy, where any snow is unusual, the red and yellow snows that season caused consternation and alarm. The world had gone awry. "During the entire season," wrote one observer, "the sun arose each morning as though in a cloud of smoke, red and rayless, shedding little light or warmth and setting at night behind a thin cloud of vapor, leaving hardly a trace of its having passed over the face of the Earth."

In New England, 1816 was called "the Year Without a Summer" and also, with Yankee wryness, "Eighteen Hundred and Froze to Death." As the spring wore on into summer, there were successive cold waves and frosts every month. We know just how cold because the presidents of Yale University during this period—a succession of clergymen and scholars—were willing to rise each morning at 4:30 to read and record the temperature. In June 1816, the average temperature was 7 degrees Fahrenheit below normal for the preceding years and those that followed. The temperature fluctuated wildly: from a low of 35 degrees at sunset on June 7 to a high of 88 at two in the afternoon on June 24. (To give credit where it is due, a similar record was being maintained at Harvard, but at this critical time the series of readings was interrupted when the keeper of the records, Samuel Williams, Hollis professor of natural history, was caught

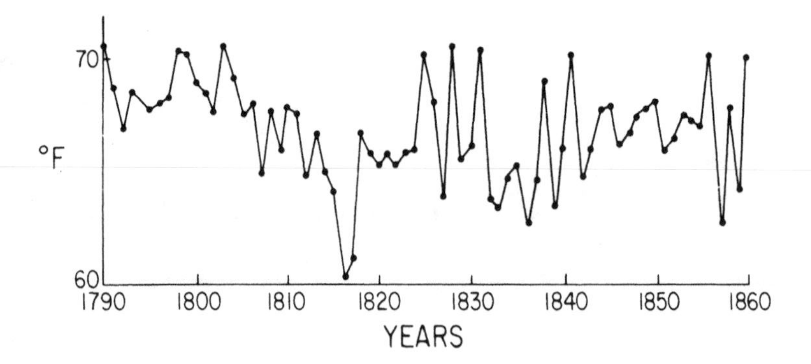

Mean temperature at New Haven, Connecticut, from 1790 to 1860. From Stommel and Stommel, 1983.

embezzling funds from the Hopkins Trust and fled to Vermont to avoid prosecution—one of the less-celebrated incidents in the continuing rivalry between Harvard and Yale.)

In New England that summer, crop failures were rife. Only a quarter of the principal staple crop, Indian corn, ripened sufficiently to be used for meal, and hay and wheat crops fared as badly. The price of corn and wheat soared by 50 percent the following winter and, there being little feed for hogs and cattle, the farmers sold off much of their stock, bringing about a corresponding collapse in the market price of beef and pork. It could have been worse: the United States was at that time, as now, an exporter of agricultural products, so there was typically a surplus beyond what was needed for local consumption. Most people adapted and survived. In fact, there were even a few benefits.

In October, the Philadelphia Society for the Promotion of Agriculture decided to collect whatever facts they could about the effects of the untimely frosts on vegetation and, among many respondents, they heard from Samuel Mitchell, a physician and professor of natural philosophy at Columbia University in New York:

> There will not be half a crop of maize on Long Island, and in the southern district of this state. Further northward there will be less. The buckwheat is so scanty, that a few days ago I paid four dollars for half a bushel, for the use of my family.
>
> The season, though unusually cool, was nevertheless warm enough to ripen strawberries, raspberries, currants,

> cherries, gooseberries, pears, plums, and apples. They were generally very fine. The ox-heart cherries, in particular, were unusually large and abundant. This autumn apples are fairer, cheaper, and more plentiful than they have been for many years. Peaches were poor, owing to the distemper of trees of several years' standing.
>
> It is certain that the fruit has been damaged less by insects than is usual. An entomologist complained to me, a few weeks ago, that it has been a most unfortunate season for the collection of insects. That kind of game, he said, was so rare, that he had added but little to his museum.
>
> There have been at New York fewer fleas and mosquitoes than ordinary.

What was a major inconvenience in America was a disaster in many parts of Western Europe, already occupied with healing itself in the aftermath of the Napoleonic Wars and far more dependent on local agricultural production. Crop failures were universal, with severe food shortages here, downright famine there. In Switzerland and France, the price of grain and bread doubled and tripled. In some districts of Germany and Switzerland, officials sealed off their borders to prevent the export of grain. By the end of the year in Ireland, the deficient harvest of the summer had been completely consumed; small landholders were forced to abandon their homes that spring and beg for a living.

"The late year," wrote a Moravian landowner in October 1816, "met with misfortune, all winter grains are still green. Nobody in my manor has any seed grain. The price is so high that it hurts! Nobody is able to buy, and also my private means is not in proportion to the enormity of need for assistance; so several thousand people become impoverished as the year progresses. Calamitous enough, that every farmer dismisses farmhands and maidservants, and because of dearth and lack of fodder sells half his cattle."

Throughout Europe, food riots broke out, armed groups raided farms, and bakeries and grain markets were looted. Landowners fortified their estates against roaming bands of the destitute. A major tenet of the Conspiration of 1817 in France held that all that was needed to lower the price of bread was the overthrow of the monarchy and a return to a Napoleonic regime. It may well be that the conditions of famine promoted the activities of

Sir Thomas Raffles (1781–1826). National Portrait Gallery, London.

the typhus-bearing louse responsible for the European epidemic of 1816–19. In Ireland alone in this period, 1.5 million people were afflicted with typhus, leading to 65,000 deaths.

It is astonishing to think that all of this disruption and human tragedy was brought about by Nature, specifically in the form of a volcanic eruption a year earlier halfway around the globe in Indonesia—the explosion of the mountain called Tambora, in April 1815.

The most definitive chronicle of this stunning event was given to the Natural History Society in Batavia (Jakarta) six months later by Sir Thomas Stamford Raffles. Raffles is considered the founder of Singapore and at the time of the eruption was British Resident in Malaya and the East Indies (Indonesia). A product of the Age of Reason, itself inheritor of the Renaissance ideal of the universal man, Raffles was an administrator but also a student of science and nature as well as philosophy and the arts. Much of our knowledge of natural events that occurred in this era derives from the careful observations of such "amateurs," and in our age of professionalization and specialization we can look back on some of these people with awe. Here is Raffles's summary of reports he received about the Tambora eruption:

Island of Sumbawa, 1815—In April, 1815, one of the most frightful eruptions recorded in history occurred in the mountain Tambora, in the island of Sumbawa. It began on the 5th day of April, and was most violent on the 11th and 12th, and did not entirely cease till July. The sound of the explosion was heard in Sumatra, at a distance of nine hundred and seventy geographical miles in a direct line, and at Ternate, in an opposite direction, at the distance of seven hundred and twenty miles.

Out of a population of twelve thousand, only twenty-six individuals survived on the island. Violent whirlwinds carried up men, horses, cattle, and whatever else came within their influence, into the air, tore up the largest trees by the roots, and covered the whole sea with floating timber. Great tracts of land were covered by lava, several streams of which, issuing from the crater of the Tambora mountain, reached the sea.

So heavy was the fall of ashes, that they broke into the Resident's house in Bima, forty miles east of the volcano, and rendered it, as well as many other dwellings in the town, uninhabitable. On the side of Java, the ashes were carried to the distance of three hundred miles, and two hundred and seventeen towards Celebes, in sufficient quantity to darken the air. The floating cinders to the westward of Sumatra formed, on the 12th of April, a mass two feet thick and several miles in extent, through which ships with difficulty forced their way.

The darkness occasioned in the daytime by the ashes in Java was so profound, that nothing equal to it was ever witnessed in the darkest night. Although this volcanic dust, when it fell, was an impalpable powder, it was of considerable weight; when compressed, a pint of it weighing twelve ounces and three quarters. Along the sea-coast of Sumbawa, and the adjacent isles, the sea rose suddenly to the height of from two to twelve feet, a great wave rushing up the estuaries, and then suddenly subsiding. Although the wind at Bima was still during the whole time, the sea rolled in upon the shore, and filled the lower parts of houses with

View of the Tambora caldera. Steven Carey, University of Rhode Island.

> water a foot deep. Every prow and boat was forced from the anchorage and driven on shore.
>
> The area over which tremulous noises and other volcanic effects extended was one thousand English miles in circumference, including the whole of the Molucca Islands, Java, a considerable portion of Celebes, Sumatra and Borneo. In the island of Amboyna, in the same month and year, the ground opened, threw out water, and closed again.

Tambora was not your garden-variety volcanic eruption, like Mount Saint Helens's in 1980. It was the largest volcanic eruption in modern historical times and one of the largest in the past 10,000 years. It ejected into the Earth's stratosphere huge quantities of dust and sulfur dioxide that before long formed a semiopaque membrane around the planet, in effect a sunscreen. Light, the visible portion of the Sun's radiation, exists in the form of relatively short wavelengths. The particles of dust and sulfur dioxide (called an *aerosol*) are about the same size as the wavelengths of the incoming light, so many of these wavelengths run into the particles and are reflected. On

Haze and greenhouse effects. Short-wavelength incoming radiation, ———. Long-wavelength outgoing radiation, ∿∿∿.

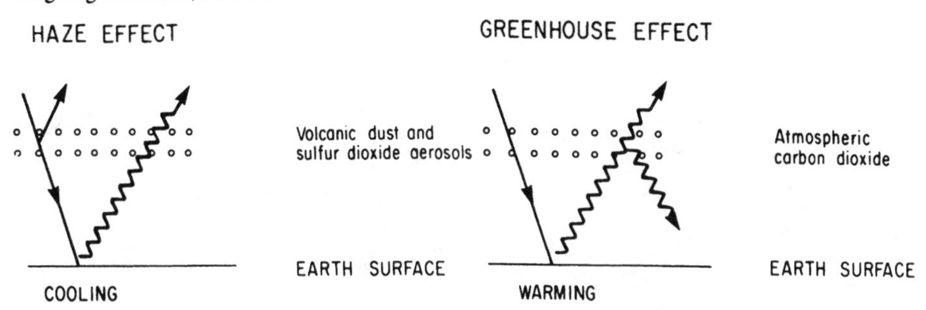

the other hand, heat reflected from the Earth is in the form of relatively *long* wavelengths, most of which don't collide with such particles but merely swing on past, escaping into space. The net effect is a cooling of the Earth; the Tambora eruption produced a reverse greenhouse effect. The "impalpable" but weighty dust particles were long thought to be the main contributor to the haze and its effects, but recent observations of smaller eruptions indicate that sulfur dioxide and sulfate particulates, which remain in the stratosphere for a year or two, are probably the more important villains.

Within a few years, the dust had all fallen, the aerosols had dissipated, and the world had returned to "normal," a relatively secure platform for life, rocked every now and then by an inexplicable global disaster like Tambora, but in general a permanent arrangement of land and sea. It had only been a few decades before Tambora that anyone had equated volcanic activity with distant climatic effects (in fact, Benjamin Franklin was the first to do so), but only in the last thirty years have we discarded the prevailing wisdom that the continents and ocean basins have always been in the same geographical locations. This was a comforting model, in a sense; it just happened to be entirely wrong. The competing hypothesis of "continental drift"—or "plate tectonics," as it is now known—has permitted us a far greater understanding of the nature of such things as volcanoes. Understanding them does nothing to tame them, of course, but it does serve to make the Earth appear a bit less whimsical in its outbursts.

In plate tectonic theory, the Earth's outer surface, or crust, is considered to be divided into a number of plates, which move horizontally at rates of a fraction of an inch to a few inches per year. New plate material is formed at

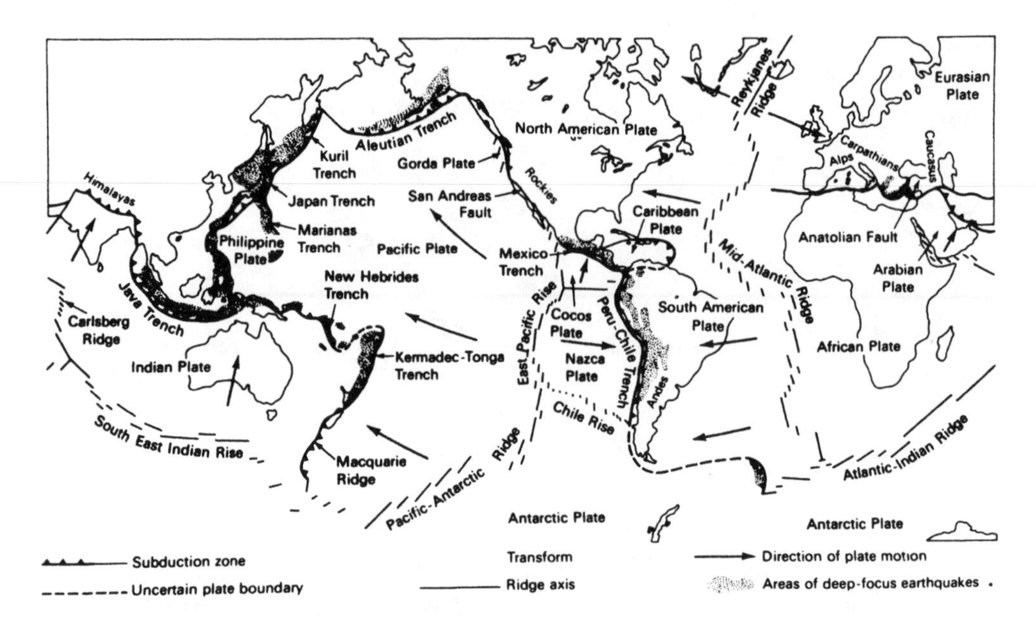

The Earth's lithosphere is broken into large, rigid plates, each moving as a distinct unit. The relative motions of the plates, assuming the African plate to be stationary, are shown by the arrows. Plate boundaries are outlined by earthquake belts. Plates separate along the axes of mid-ocean ridges, slide past each other along transform faults, and collide as subduction zones. From Uyeda, 1971.

their originating ends and old plate material is subducted back into the Earth at their trailing ends. The new plate material consists of molten magma which has been brought up from depth, particularly along the mid-ocean volcanic ridge system. The old plate material is carried back down into the mantle of the Earth, principally along the major earthquake zones surrounding the Pacific Ocean. The plates themselves move as rigid slabs over the viscous and underlying mantle and are considered to be driven by thermal convection currents in the mantle.

That the continents may have drifted about on the Earth's surface is an idea often attributed to Francis Bacon, essayist, lord chancellor to James I, candidate for those who don't believe Shakespeare wrote his own plays, and later subject of a bribery conviction. While computer analyses and other studies have shown that the author of Shakespeare's plays was almost certainly a man named William Shakespeare, a consultation of Bacon's own writings shows that while he did note in 1620 the obvious similarities of the continental outlines on either side of the Atlantic Ocean, he did not suggest that at one time they might have formed a unified land mass. That possibility was espoused in *The Origin of Continents and Oceans*, an elegant book by German meteorologist Alfred Wegener first published in German in 1912 and translated into English in 1915. The idea was dismissed by most Earth scientists as inept and unscientific. After all, it challenged the very fundament of geologic thinking and, if accepted, would have called for a wholesale rethinking of how the Earth works.

Wegener was present at a meeting in 1926 of the American Association of Petroleum Geologists during which a symposium was held on the subject. Or perhaps it might be more aptly called an ambush. A professor from the University of Chicago commented that geologists might well ask if theirs could still be regarded as a science when it is "possible for such a theory as this to run wild." Another from Johns Hopkins University commented on Wegener's methodology: "It is not scientific but takes the familiar course of an initial idea, a selective search through the literature for corroborative evidence, ignoring most of the facts that are opposed to the idea, and ending in a state of autointoxication in which the subjective idea comes to be considered as an objective fact." In these words, not only Wegener's hypothesis but Wegener himself was under attack. Contrary to a common perception of scientists as dignified, objective investigators, they often play hardball with subjective zeal—especially when their basic premises are challenged.

Widely considered a pseudoscientific notion, the matter of continental drift rested until the 1960s. A sudden change in attitude toward the matter is generally attributed to the publication in 1963 of a scientific article by Fred Vine and Drum Matthews of Cambridge University. It was well known by then that the Earth acts like a great magnet that switches its magnetic polarity through geologic time. Sometimes, in essence, the North Pole becomes the South Pole, and vice versa. And as molten rock cools and hardens, magnetic particles in the lava are "frozen" like little compass needles in the hard rock, their direction depending on the state of the Earth's magnetic polarity at the time. Vine and Matthews took note of the alternating positive and negative magnetic "stripes" in the rock, which parallel the great ridges that occur on the mid-Atlantic sea floor, and suggested that they could best be explained if the sea floor itself were spreading out—moving away from the ridges. The particular magnetic signature would be picked up as the molten lava cooled, the signatures alternating in stripes as the Earth's polarity switched back and forth through the geologic ages. Thus, however dimly perceived, there was now a mechanism by which the continents might indeed have spread.

Evidence in favor of this hypothesis began to cascade, and, in spite of a few serious contrarians and the waggish carping of a group that called itself the Stop Continental Drift Society, continental drift is now accepted as the explanation for the present configuration of the continents and is considered a major feature of the Earth's continuing metabolism.

It is not irrelevant, however, to note that prior to the publication of Vine and Matthews's article, Lawrence Morley of the Canadian Geological Survey had submitted a manuscript with a nearly identical but independently derived explanation for the striped pattern of magnetic "anomalies" on the sea floor. His manuscript was submitted to two leading scientific journals, went through the usual peer review system, and was turned down. One reviewer commented that such a subject was suitable only for cocktail conversation.

The term *peer review*, which is the system most scientific journals employ to determine the acceptability of submitted manuscripts, has a benevolent aura to it that is not always warranted. Given the vast quantity of manuscripts seeking publication in a limited number of journals, editors need ready reasons for rejecting them, and one bad review is generally sufficient, thus relieving the glut. A good number of unworthy manuscripts are rejected in this way, but one can also be fairly sure that a manuscript that is innovative or speculative, or that challenges the common wisdom, will similarly be thrown out. Fortunately, there are some experienced editors who appreciate this problem and treat with extra care those manuscripts they think might contain a nugget that could lead to extended scientific advances.

The same peer review system applies to proposals made for research funds, at least those submitted to the federal government—and the same fate usually awaits proposals for innovative research programs. A number of mature scientists have developed the technique of cheating a little on the system: in their research proposals, they describe work they have just completed with success, and once their "proposal" is funded, they set out to do what they wanted to do in the first place.

There is an alternative to the peer review system, practiced more often in Canada and Europe than in the United States, but it lacks democracy. In this system, the funding organization poses the scientific questions that appear to be pertinent and then picks the scientists who have the ability and interest to do the research. The problem here is that you may have picked the wrong race *or* the wrong horses. In either system, it is hard for society at large, as well as the great run of scientists, to know what has been missed.

In any event, in spite of the politics inherent in science and other endeavors, and in spite of other birth pangs, the plate tectonics model of Earth came into being and soon explained much of the nature of the planet's being, including volcanism.

Volcanism is directly associated with the mid-ocean ridges, where molten material fills the gaps that occur as the sea floor spreads. Such spreading occurs as two "plates" move in opposite directions from the ridge. The ridges are thought to be fed, either directly or indirectly, with molten material that comes up as giant plumes from a great depth. Called *mantle plumes*, they are presumed to originate near the boundary between the lower mantle and the liquid core of the Earth, about halfway to the center of the Earth. The rising molten material in these plumes is basic in composition (as opposed to acidic), dominated by heavy minerals, and enriched in sulfur dioxide, which, along with the contained carbon dioxide, chlorine, and water, is vaporized when the molten magma erupts at the Earth's surface. These latter components of the magma are called its volatile constituents.

It stands to reason that if two plates are moving away from each other in one place, they will be crashing against something else at the other, leading end, and this is what happens. In many cases, where two plates collide, there is a subduction zone, where one or both of them descend back into the mantle. When the edge of a plate is subducted to a sufficient depth, its material reaches temperatures high enough to bring about at least partial melting, which in turn produces chambers of magma that tend to rise up. Volcanism then occurs at the Earth's surface. Volcanoes that occur in subduction zones (such as Indonesia) typically spew out more acid debris composed of lighter materials from the subsumed and overlying crustal materials. One such group of volcanoes circumscribing the Pacific Ocean is known as the Ring of Fire.

One of the mantle plumes that feeds the mid-Atlantic ridge rises in the North Atlantic directly under the island of Iceland, which can be thought of as the child of this plume, and is one of the several visible parts of the ridge, others being the Azores, Saint Peter and Saint Paul Rocks, and Tristan de Cunha. On Iceland, a volcano called Laki erupted with a gigantic lava flow in June 1783, and the eruption continued for eight months, a dramatic example of volcanic pollution. An enormous amount of sulfur dioxide was ejected into the atmosphere, returning to Earth as an estimated 100 million tons of sulfuric acid rain. This is about the same amount of acid rain attributable to human causes that today falls on the Earth in an entire year.

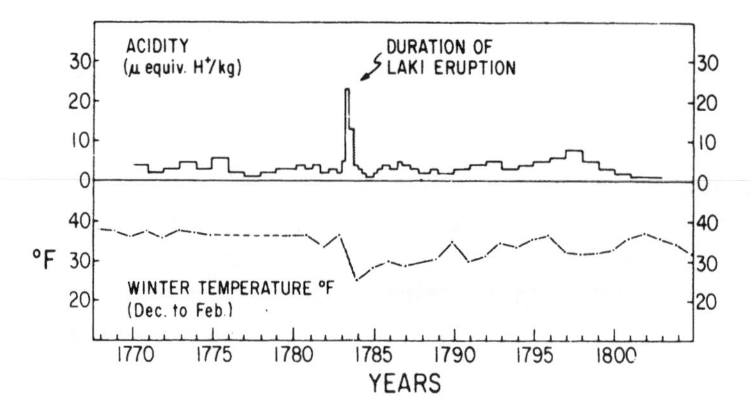

The effects of the Laki eruption in 1783, as reflected by acidity levels of precipitation on the Greenland ice sheet and the winter (December to February) temperature record of the eastern United States. From Sigurdsson, 1982.

Happily for modern scientists, there is a record of this and other such events, a record as precise as the annual growth rings in trees but going further back than any living tree. In the more northerly latitudes of Greenland, snow and ice deposition increases in layers year by year, and the effects of a great variety of unusual atmospheric events become trapped in these layers, accessible by means of ice cores. Peaks of high acidity from the sulfur dioxide aerosols that have settled back to Earth in the ice cores have now been correlated with all known volcanic eruptions, and the Laki eruption in 1783 created higher-acidity peaks than any other volcanic eruption in the past thousand years. But there were more immediate effects. As at Tambora, the sulfate aerosols produced regional cooling; it was about 7 degrees cooler in eastern North America during the following winter, with normal temperatures returning during the next two to three years. The European winter was unusually harsh. Thanks to the sulfur dioxide in the stratosphere, a bluish haze spread over Iceland and all of Europe, reaching as far east as Siberia and as far south as North Africa. Acid rain proceeded to destroy most of Iceland's crops and 75 percent of its livestock, its effects reaching England and Norway.

At that time, Benjamin Franklin was the American ambassador to France. In his seventies and, like Raffles, a man of extraordinarily broad interests and accomplishments, he had recently helped negotiate the preliminary peace treaty to end the American Revolution and would sign the final

Benjamin Franklin (1706–1790). Metropolitan Museum of Art, New York.

one, eventually to become the only person to sign all four of the key documents of American nationhood: the Declaration of Independence, the treaty of alliance with France, the peace treaty itself, and later the Constitution. Nevertheless, this indefatigable man took time out from his ambassadorial duties to ponder the bizarre climatic conditions that beseiged Europe and America in 1783 and 1784. He presented his conclusions before the Philosophical Society of Manchester in the latter year at the age of seventy-eight.

> During several of the summer months of the year 1783, when the effects of the sun's rays to heat the earth in these northern regions should have been the greatest, there existed a constant fog over all Europe, and a great part of North America. The fog was of a permanent nature; it was dry, and the rays of the sun seemed to have little effect towards dissipating it, as they easily do a moist fog, arising from water.

> They were indeed rendered so faint in passing through it, that when collected in the focus of a burning glass, they would scarce kindle brown paper. Of course, their summer effect in heating the earth was exceedingly diminished.
>
> Hence the surface was early frozen.
>
> Hence the first snows remained on it unmelted, and received continual additions.
>
> Hence the air was more chilled, and the winds more severely cold.
>
> Hence perhaps the winter of 1783–4, was more severe, than any that happed for many years.
>
> The cause of this universal fog is not yet ascertained. Whether it was adventitious to this earth, and merely a smoke, proceeding from the consumption by fire of some of those great burning balls or globes we happen to meet with in our rapid course round the sun, and which are sometimes seen to kindle and be destroyed in passing our atmosphere, and whose smoke might be attracted and retained by our earth; or whether it was the vast quantity of smoke, long continuing to issue during the summer from Hecla in Iceland, and that other volcano which arose out of the sea near that island, which smoke might be spread by various winds, over the northern part of the world, is yet uncertain.
>
> It seems however worth the enquiry, whether other hard winters, recorded in history, were preceded by similar permanent and widely extended summer fogs. Because, if found to be so, men might from such fogs conjecture the probability of a succeeding hard winter, and of the damage to be expected by the breaking up of frozen rivers in the spring; and take such measures as are possible and practicable, to secure themselves and effects from the mischiefs that attended the last.

Here then is the first known suggestion of a link between volcanoes and widespread atmospheric effects, a link that was well understood when, almost exactly 100 years later in 1883, sailors in the Sunda Straits between Java and Sumatra suddenly lost one of their chief aids to navigation, the island called Krakatoa. Unlike the leisurely, eight-month eruption of Laki,

Krakatoa was a violent and abrupt series of events. There had been a few precursor eruptions, small ones extending from May through July 1883. But the major eruptions took place on two successive days in August, the 26th and 27th, and nearly obliterated the island, which was uninhabited at the time. By the standard set by nearby Tambora sixty-eight years earlier, Krakatoa was minor, but such comparisons mean little to people whose loved ones lose their lives in such events. And the damage was considerable—not necessarily from the eruption itself but from the train of events it set in motion: specifically, a tsunami, or giant sea wave, that swept away more than 30,000 people on the islands of Java and Sumatra. Commonly called tidal waves, these low walls of water sweeping across the ocean toward land are caused typically by the sudden vertical movement of the sea floor, often the result of earthquakes near deep sea trenches. In the cool parlance of physics, they propagate across the ocean as a low-amplitude and long-wavelength disturbance, but as they approach shallow waters near land, their amplitude increases; in other words, the tsunami rises up as a rushing wall of water ten feet or more high. The tsunami that swept across the water after Krakatoa erupted was either the result of ejected material plummeting into the sea or the collapse of the island structure after the eruption. One of the few survivors living in the town of Anjer in Java saw it this way:

> At first sight, it seemed like a low range of hills rising out of the water, but I knew there was nothing of the kind in that part of the Sunda Strait. A second glance—and a very hurried one it was—convinced me that it was a lofty ridge of water many feet high. . . . There was no time to give any warning, and so I turned and ran for my life. My running days have long gone by, but you may be sure that I did my best. In a few minutes, I heard the water with a loud roar break upon the shore. Everything was engulfed. Another glance around showed the houses being swept away, and the trees thrown down on every side. Breathless and exhausted I still pressed on. . . . A few yards more brought me to some rising ground, and here the torrent of water overtook me. I gave up all for lost. . . . I was soon taken off my feet and borne inland by the force of the resistless mass. I remember nothing more until a violent blow revived me. . . . I found myself clinging to a coconut palm. Most of the trees near

> the town were uprooted and thrown down for miles, but this one fortunately had escaped and myself with it.... As I clung to the palm-tree, wet and exhausted, there floated past the dead bodies of many a friend and neighbour. Only a mere handful of the population escaped.

By January 1884, it was obvious that the eruption had created global atmospheric effects and the Royal Society of London appointed a committee to look into these phenomena. Krakatoa's renown comes not from the damage it caused, as tragic as that was, but rather from the fact that it was the first major geophysical event of any kind to be studied scientifically on a global scale. A good number of such events have been studied since in such a manner, but none that had the immense potency of Krakatoa.

Four years later, the Royal Society released its report, the result of data collected from well over fifty observatories around the world. To begin with, it was one of the loudest sounds in history, the eruptions being heard as far as 3,000 miles away—farther than the distance between San Francisco and New York. In New Guinea, 1,800 miles from the island, Dr. F. H. Guillemard reported that "the Rajah of Salwatty, whom I met at the village of Samatu, told me that the noise of the eruption had been audible there." More than 2,000 miles away, at Alice Springs in South Australia, a Mr. Skinner reported that "two distinct reports, similar to the discharge of a rifle, were heard the morning of the 27th, and similar sounds were heard at a sheep camp nine miles west of the station, and also at Undoolga, 25 miles east." And 3,000 miles away, on the island of Rodriguez in the Indian Ocean, chief of police James Wallis reported that "several times during the night of the 26th–27th reports were heard coming from the eastward, like the distant roar of heavy guns."

Low-frequency sound waves circled the earth as many as seven times in the course of five days after the eruption, picked up by barographs (instruments for measuring barometric pressure) at stations all around the Earth. For a year following the eruption, floating pumice was found throughout the Indian Ocean, a phenomenon noted in many a ship's log. From January 5 to 10, 1884, Captain Clarke of the schooner *Lord Tredegar* was plying the western Indian Ocean some 2,900 miles from Krakatoa and encountered huge rafts of pumice. In March of that year, Captain Gray of the ship *Parthenope* found the central Indian Ocean strewn with pumice that was covered with barnacles, testimony that it had been in the water a long time.

The stuff washed up on the shores of Durban, South Africa, on September 27 of that year, and ships' reports of floating pumice continued into November.

More long lasting were the eruption's atmospheric effects. For three years, in many parts of the world, the days were filled with a blue or green haze, with spectacular red glows arising just after sunset and just before dawn. (More recent studies have shown, not surprisingly, that these aerosol effects were accompanied by a global cooling of about one degree, with temperatures returning to normal in the same three-year period.) In Poughkeepsie, New York, there was such an "intense glow in the sky that fire engines were called in the morning" on November 27, 1884. "As soon as the sun's disc has disappeared," wrote a Mr. Layard halfway around the world from Poughkeepsie in Noumea, New Caledonia,

> a glow comes up from the west like that of white-hot steel, reddening somewhat as it mounts to the zenith, but changing the while to blue. From the zenith it passes into the most exquisite green, deepening as it loses itself in the east. As the sun sinks lower and lower, the red tints overpower the white-hot steel tints, and the blue of the zenith those of the green. At 7 p.m., or a little after, nearly the entire western half of the horizon has changed to a fiery crimson: as time goes on, the northern and southern areas lose their glory, and the greys of night contract, from the northern end first, most rapidly; the east is of a normal grey. The south now closes in, and presently, about 8 p.m., there is only a glare in the sky, just over the sun's path, as if a distant conflagration, till the fire in the west dies out. I have been attempting to describe one of the cloudless evenings, of which we have had only too many, having just come through a fearful drought that has lasted all this while; but who shall paint the glory of the heavens when flecked with clouds?—burnished gold, copper, brass, silver, such as Turner in his wildest dreams never saw, and of such fantastic forms!

To the writer's poetic attempts to describe these skies, and to any artist's inspired versions, the scientist can but add what may seem a few mundane stage directions to accompany the role played by the stratospheric layer of volcanic dust and aerosols, that of a gigantic projection screen in the sky, illuminated by the Sun's rays.

The human eye is sensitive to only part of the electromagnetic spectrum emitted from the Sun—the part we call visible, or white, light. If white light is shone through a prism, each wavelength among those that make up light is bent (or *refracted*) to a slightly different degree and shows up, by our color-coded definitions, as blue to green to yellow to orange to red light. The wavelengths embodied in white light vary from 0.48 microns for blue to 0.68 microns for red (a micron equals a millionth of a meter). Blue light is short wavelength, red is longer. Even shorter than blue is ultraviolet (light) and longer than red is infrared (heat), neither of which is visible.

The particles in the stratospheric volcanic dust and aerosol layer, as well as those in the atmosphere below, are in the range of 0.5 to 2 microns. So, for blue light, these particles will appear fairly large: there will be a lot of "collisions" and a substantial scattering of blue light. For red wavelengths, the particles will appear smaller: fewer collisions and less scattering. The effect is that as sunlight passes through these particles, the blue light is scattered and largely lost to our view. The light that escapes unscathed will thus be "weighted" toward the red end of the spectrum. The longer the passage through the atmosphere, the more reddened will be the light that we see. Hence red sunsets: the rays of the sun are passing through the atmosphere sideways (as far as we are concerned), rather than vertically through the atmosphere as at noon, so more of the blue is scattered during the journey. A volcanic layer of particles amplifies and alters this effect.

Picture a sunset. To the west, a layer of volcanic dust is illuminated from above by direct sunlight (which has not been reddened by passage through the atmosphere). The volcanic layer appears silver with a bluish tinge, the result of scattering by the layer itself. Near the zenith, overhead, scattered blue light predominates. But to the east, the layer is illuminated from below (or inside), and the Sun's rays have had a long, oblique passage through the atmosphere. The light reflected by the volcanic layer in the east is pink.

It is now twenty minutes after sunset. The silver portion of the western sky diminishes and is followed by a dark band, the result of rays that have had a long passage through the dense volcanic layer with a substantial reduction of light of all wavelengths. Toward the east, the volcanic layer is illuminated from below: the colors range from pink to red, corresponding to progressively longer passages of the Sun's rays through the atmosphere. And, to the east, a gray region appears because there is no solar illumination

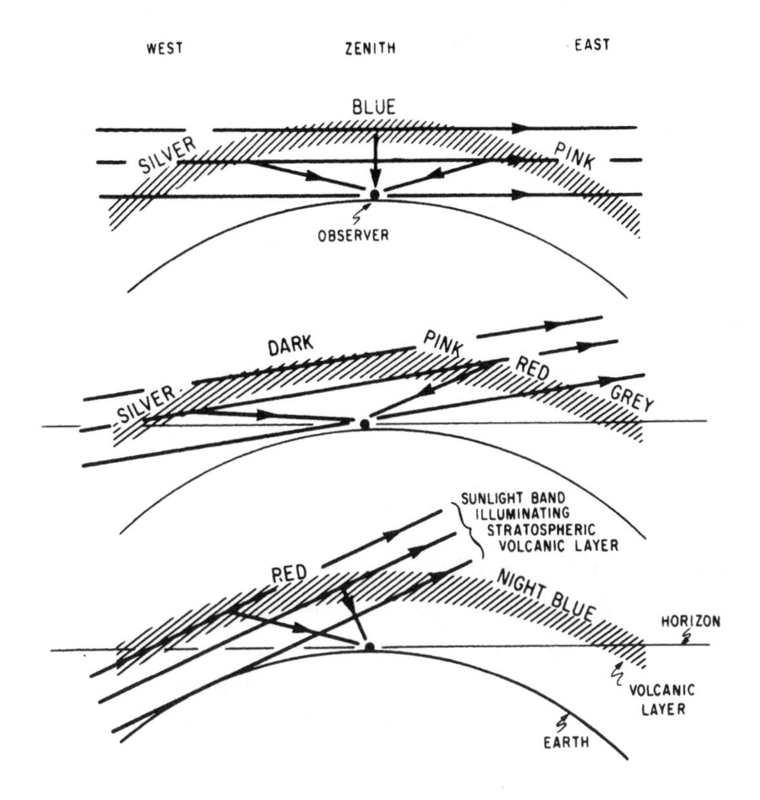

Geometry of sky display following the eruption of Krakatoa. The three views are, sequentially, for sunset, twenty minutes after sunset, and forty minutes after sunset. Adapted from Meinel and Meinel, 1983.

of that part of the volcanic layer. Forty minutes after sunset, the only display is the brilliant red.

The sequence of events in a sunrise is simply reversed.

As violent and terrifying as they are, most volcanic eruptions do not seem to have changed the course of history to any great extent. The truly grand eruptions have occurred before humans were present, and most of the big ones since we arrived have occurred in places that were relatively remote and uninhabited. Even so, regional effects can be and have been devastating on human populations. Vesuvius simply stopped life in Pompeii in its tracks in A.D. 79, a tragedy for the inhabitants though a bittersweet boon to archaeologists. But life went on largely as before in the nearby areas. Volcanic eruptions around 1,000 years ago in northern Arizona forced the farming people we call the Sinagua from the region, but they were given a new lease

Dates, Locations, and Magnitudes of Selected Volcanic Eruptions

Name	Year	Country	Estimated Ejecta Volume (cubic kilometers)
Toba	73,000 B.C.	Indonesia	1,000
Santorini	1470 B.C.	Greece	10
Vesuvius	A.D. 79	Italy	1
Laki	A.D. 1783	Iceland	10
Tambora	A.D. 1815	Indonesia	100
Krakatoa	A.D. 1883	Indonesia	10
Saint Helens	A.D. 1980	United States	1

Adapted from Stommel and Stommel, 1983.

on life: the volcanic ash enriched the arid soils and in due course the people returned to better harvests. What would happen if a massive volcanic eruption occurred in the midst of a crowded civilization? We know the answer because just that happened around 1470 B.C. While volcanoes have given tribal people entire mythologies if not religions—such as the Hawaiians—this eruption changed the course of Western civilization almost overnight. It also provided the West with one of its most enduring, even tenacious, legends: Atlantis.

In 1470 B.C., a volcano we today call Santorini, located on the small island of Thera a few miles north of Crete in the eastern Mediterranean, exploded in a sequence of eruptions that, in all, were probably the equivalent of Krakatoa. This was the proximal center of Western civilization at the time, specifically, Greece, Crete, Asia Minor, and Egypt. There is no written history of these events—only legends written many years later, recent archaeological findings, and the geologic record.

It is perfectly evident that the island of Thera was devastated by a volcanic eruption sometime in the past because it is covered with volcanic ash, indeed, with four distinct layers related to four distinct phases of the overall eruption. The first and second phases (which were smaller in magnitude than the later phases) deposited a layer that is up to twelve feet thick on the island, the second depositing a layer up to nearly twenty-four feet thick. Between these two phases and the next two, there was a considerable time interval. We know this because the boundary between the second and third layers is sharp, often showing erosional effects on the second layer. In some locations there are gravel deposits and layers of soil between the two layers, suggesting

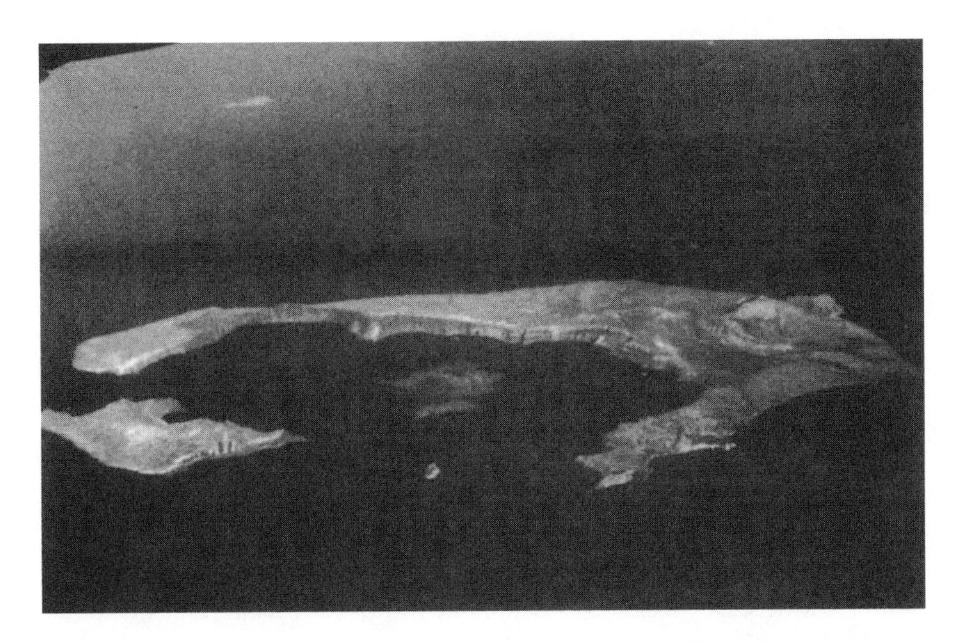

Aerial view of the island of Thera and its caldera. The small islands in the middle of the bay are the present sites of volcanic activity. From Luce, 1969.

an interval of ten to fifty years between the first two phases of eruption and the third.

There was a considerable settlement on Thera at the time. It was an outpost of the Minoan civilization that flourished on Crete, and the first two eruptions, though nothing to match what was in store, sealed the Therans' fate as tightly as Vesuvius sealed Pompeii. The remains of the settlement were found serendipitously in 1866, as a result of the need for volcanic pumice to mix with lime to make a durable and seawater-resistant cement for the construction of the Suez Canal. Pumice quarriers found the ruins of the ancient settlement, and investigators soon determined that it was Minoan in character. Archaeological artifacts allowed the destruction of Thera to be dated: radiocarbon dates for wood and fava beans gave mixed results (perhaps because of contamination) of between 1600 and 1400 B.C. Pottery seems more reliable in this case; the last datable vase is comparable in style to that of the Late Minoan IA period on Crete, with a generally accepted end date of about 1500 B.C. This would put the main phases of eruption—the third and fourth—some thirty years later, or about 1470 B.C.

The inhabitants probably had warning that their island was about to erupt, since archaeologists found no skeletons and few portable objects that were of any particular contemporary value. The Therans may have come back briefly between the first two phases, but the evidence is that after the second eruption they abandoned the place for good.

The third and fourth eruptions deposited layers of ash on the island that were, respectively, 180 and 130 feet deep in places. It has been estimated that the third phase accounts for 57 percent of all the volcanic debris on the island. Deep sea coring in the eastern Mediterranean has added to this violent legacy: a layer of ash in the sediments correlates precisely with the Santorini eruptions, extending in a southeasterly direction away from Thera, in accord with the direction of the prevailing winds. On Crete itself, seventy miles away, the ash ranges from one-half inch to two inches thick. Comparing the details of this ash layer with that of the eruptions of Krakatoa and Laki, whose other effects are fairly well documented, it is possible to guess with some accuracy what happened in the rest of the Mediterranean and its civilizations when Santorini erupted.

The major eruption on Krakatoa lasted for two days and its eruption column rose to a height of thirty to fifty miles. Such a column on Thera would have been visible on a clear day from Egypt, some 450 miles away. The main destructive agent from the Krakatoa eruption was the resulting tsunami, which rose to a height of fifty feet or more at distances of forty miles from its origin. Presumably the collapse of the Santorini caldera, the large basin-shaped depression left after the volcanic eruption, would have produced much the same effect—a tsunami sufficient to devastate coastal settlement on the northern and eastern shores of Crete, and sufficient to be felt on the coast of Egypt. The ash that fell on Crete would have ruined any crops growing at the time and poisoned the soils for years to come. And, as with Laki, acid rain would have worsened the Cretan agricultural situation. This multiple devastation is what is inferred for Crete from the geologic evidence.

The archaeological story of Crete may be summarized as follows: throughout much of the Bronze Age, the Minoan civilization on Crete was preeminent, producing exquisite art and architecture, including numerous palaces, the most famous of which is at Knossos. Crete controlled trade routes in the Aegean and maintained island colonies as well as outposts in Asia Minor. But in what archaeologists call the Late Minoan IB period, something happened that brought the Minoans to their knees. Their fleets were destroyed,

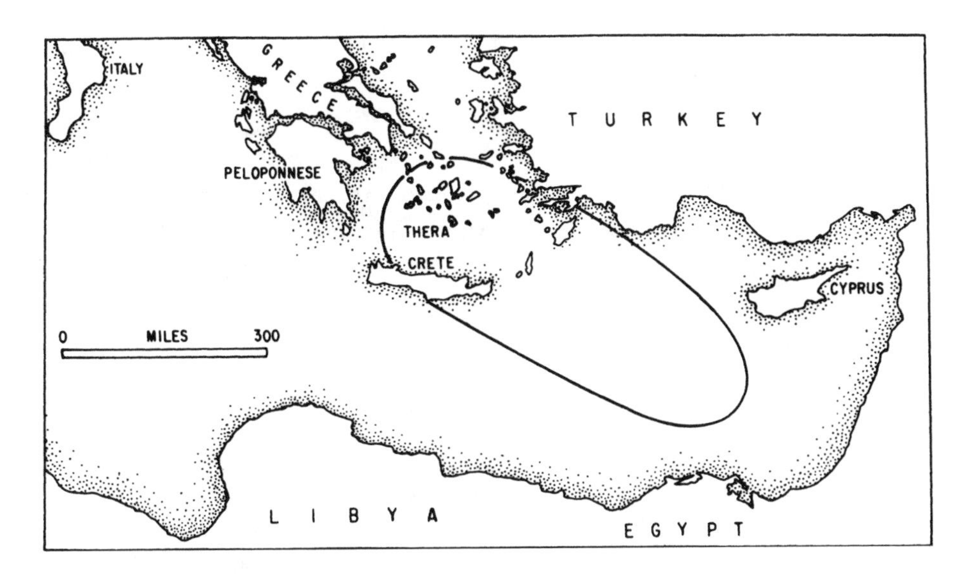

The eastern Mediterranean, showing the islands of Thera and Crete and the distribution of the volcanic ashes from the Santorini eruptions. Adapted from Luce, 1969.

along with most of their palaces and the coastal settlements in the east and north, never to be reoccupied. Trading posts overseas were abandoned. The balance of power shifted from Crete to Greece, where the Mycenaean civilization was on the rise. Greeks occupied the palace of Knossos, which had been built on high ground. Archaeologists puzzled over this rapid decline that brought to an end the Minoan IB period, dated to 1500–1450 B.C. It did not seem that Crete had been overrun by a foreign power or famine or disease. There was no plausible explanation until the Santorini eruptions were pinned down. It is now generally accepted that the entire civilization had met its demise at the hands of a volcano. It is also now generally accepted as the seed that gave rise to the Atlantis legend.

Readers of Garry Trudeau's cartoon "Doonesbury" are familiar with Hunk-ra, the ancient Atlantean prince/warrior who speaks—usually in uncontrollable fits of rage—through the medium of the fey actress Boopsie. Hunk-ra may be the most famous denizen of Atlantis today, but there are many people who are convinced that there was an actual Atlantis, a vast continent where civilization rose to heights never achieved again, where mind and technology were nearly paradisiacal, and that still speaks to the chosen today, often through the powers of quartz crystals.

This all appears to be a great deal of poetic license taken over the ages with a story embellished beyond a few actual facts by one of the great storytellers of all time, the philosopher Plato. The first known mention of Atlantis is found in his narratives *Timaeus* and *Critias*, written around 350 B.C. Plato has one of his interlocutors explain that the story was passed down to him through several generations, having originated with a Greek wise man named Solon, who in turn had heard it from an Egyptian priest. According to the account, the Egyptian was rather amused that this Greek, Solon, knew so little about the past, and went on to instruct him about it.

> And so you lived under laws like these and under even better laws, surpassing all men in every virtue, as was only right for those who had been produced and reared by the gods. Thus it is that there are many great actions of your city written down here that are marvelled at. And of all these one stands out because of its magnitude and because of the courage it involved. For the records show how your city once stopped a great power as it was arrogantly spreading itself over the whole of Europe and Asia together, after starting from the Atlantic Ocean outside. For at the time the sea was navigable there, and there was an island in front of the mouth which you Greeks call the Pillars of Heracles. The island was larger than Libya and Asia taken together, and *from it there was access to the other islands for those making the trip at this time. From these islands they could go right to the continent opposite, the one that encircles the real sea*. For all of this that lies within the mouth which I am talking about seems to be a kind of harbor with a narrow entrance. But that outside is really an ocean and the land surrounding it may be quite correctly called a continent.
>
> *Now on the island of Atlantis there arose a great and marvelous power with kings ruling over all the island, as well as many of the other islands and parts of the continent.* What is more, of these lands here inside they ruled over Libya as far as Egypt and Europe as far as central Italy. At one point, then, this power, completely united, tried with one assault to enslave your country, our country, and all the territory within the strait. Then it was, Solon, that the

Plato. Staatliche Museum, Berlin.

power of your city stood out for all men because of the courage and strength that she showed. For she surpassed all in valor and expertise in war, serving on the one hand as the leader of the Greeks and on the other hand standing alone out of necessity when the others deserted her. After experiencing extreme danger, she defeated the invaders and set up a memorial to her victory. Thus she prevented those who had not yet been enslaved from being made slaves and ungrudgingly set free the others of us who live within the boundaries of Heracles.

But at a later time there occurred violent earthquakes and floods and one terrible day and night came when your

> *fighting force all at once vanished beneath the earth and the island of Atlantis in similar fashion disappeared beneath the sea.* And for this reason even now the sea there has become unnavigable and unsearchable, blocked as it is by the mud shallows which the island produced as it sank [emphasis added].

Here in three paragraphs can be found most of the vagaries and inconsistencies that have plagued interpretation of the Atlantis legend from the time of Aristotle (Plato's student, who declared the legend nonsense) to today's seekers of cosmic power centers. The priest had confused Minoan and Mycenaean civilization, along with later Greek civilization and the city of Athens. (On the other hand, these places were all to the north of Egypt, a sedentary civilization that knew of Greeks only because they visited Egypt.) In fact, no one in Plato's time (or Solon's) had any historical record of their Mycenaean predecessors, who were only recently rediscovered.

If the first of Plato's quoted paragraphs gives an inflated idea of the size and area of Atlantis, however, the italicized phrases suggest Crete. And the phrases in italics in the second paragraph accurately describe the political status and extent of Minoan civilization. Finally, in the third paragraph, the fabled end of Atlantis accords extraordinarily well with the geologic evidence from the Santorini eruptions.

The circumstantial evidence is compelling. More circumstantial evidence can be found in another enduring story of Western civilization: the flight of the Jews from Egypt under the leadership of Moses.

Favored as scribes and in other professional categories of service, the Jews' status depended on the whim of the particular pharaoh at the time. Jewish migrations to and from Egypt took place over a period of about 2,000 years, continuing until at least the time of Christ with the flight of the Holy Family into Egypt. The exodus led by Moses has been variously dated, from about 1500 B.C. to about 1200 B.C., but it may be that the book of Exodus reports, in one synthetic tale, the stories of several such departures over the years. According to Exodus 13:20–22: "After leaving Succoth they camped at Etham on the edge of the desert. By day the Lord went ahead of them in a pillar of cloud to guide them on their way and by night in a pillar of fire to give them light, so that they could travel by day or night. Neither the pillar of cloud by day or the pillar of fire by night left its place in front of the people."

That is a remarkably apt description of what people on the coast of Egypt would have seen during the main phase of the Santorini eruptions. The account goes on, Exodus 14:21–28, as Moses and his followers reached the Sea of Reeds, a marshy coastal region now occupied by the Suez Canal:

> Then Moses stretched out his hand over the sea, and all that night the Lord drove the sea back with a strong east wind and turned it into dry land. The waters were divided, and the Israelites went through the sea on dry ground, with a wall of water on their right and on their left.
>
> The Egyptians pursued them, and all Pharaoh's horses and chariots and horsemen followed them to the sea. During the last watch of the night the Lord looked down from the pillar of fire and cloud at the Egyptian army and threw it into confusion. He made the wheels of the chariots come off so that they had difficulty driving. And the Egyptians said, "Let's get away from the Israelites. The Lord is fighting for them against Egypt."
>
> Then the Lord said to Moses, "Stretch out your hand over the sea so that the waters may flow back over the Egyptians and their chariots and horsemen." Moses stretched out his hand over the sea, and at daybreak the sea went back to its place. The Egyptians were fleeing from it, and the Lord swept them into the sea. The water flowed back and covered the chariots and horsemen—the entire army of Pharaoh that had followed the Israelites into the sea. Not one of them survived.

This sounds remarkably similar to the effects of the Krakatoa tsunami. The Santorini tsunami, in its initial phase (when the Santorini caldera collapsed), would cause the withdrawal of the sea from the Egyptian coast and its marshlands. An hour or two later, the main positive (or influx) stage of the tsunami would have plowed into the Egyptian coast just as it had earlier into Crete.

This is a highly tenuous suggestion, and it is by no means intended to preempt the Lord's role in the departure of Moses from Egypt. The hand of God may simply be one step removed, in the sense of employing natural forces to accomplish His ends. If anything, it would seem that a natural event, and one that had major effects elsewhere in the civilized world of the

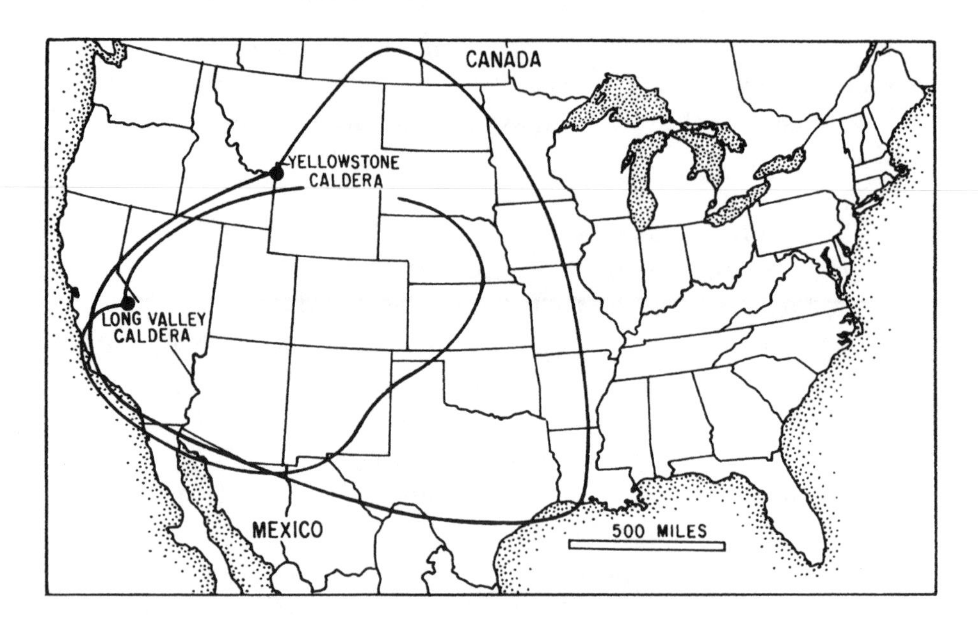

Distribution of volcanic ashes across western North America from the Yellowstone and Long Valley eruptions. From Francis, 1983.

Mediterranean, adds credence to the story as given in the book of Exodus. Certainly, no such gigantic event would have gone unnoticed.

Far more colossal eruptions have occurred. The main phase of the Santorini eruptions was ten times larger than the recent Mount Saint Helens event. The Tambora eruption was, in turn, ten times larger than Santorini. Even more enormous eruptions have taken place earlier. In Sumatra, about 75,000 years ago, the Toba eruption left a collapse structure, or caldera, that was thirty by sixty miles in dimension. Ash was spewed over much of the Indian Ocean, leaving a layer on the bottom of up to four inches thick at distances of 1,300 miles away. An eruption of the same size occurred at Yellowstone National Park in the United States 600,000 years ago. The Old Faithful geyser, the current ground movement, and swarms of minor earthquakes all suggest that it could happen again at some point in the future. The Yellowstone eruption spread ashes over most of the United States west of the Mississippi and may well have wiped out all of the vegetation in that vast region, as well as creating havoc on a global scale.

2

. . . And from Time to Time Its Surface Moves Around

Well water grew turbid and hibernating snakes took up their warm-weather rounds. Groundwater levels rose and fell. Radon levels changed radically and sensors showed that the ground was tilting. Domestic animals behaved strangely. A report notes that "pigs bit each other and tried to run up walls, cows fought each other and pawed the ground, deer ran away, turtles were seen to jump out of the water and to make noise, and a hen was seen to fly to a tree top."

This took place in Liaoning Province in northeastern China in early February 1975. Earlier, in January, the Earthquake Research Branch of the provincial government had predicted that a major earthquake would occur in the region that month or the following month. A massive quake had occurred in nearby Bo Hai in 1969, and Liaoning Province had been laced with monitoring devices in the intervening years.

By late January, the signs were growing more intense. Some wells stopped flowing altogether; others spouted water. The electrical conductivity of the ground itself changed. All of these signs were considered earthquake "precursor anomalies." The most important of them all was a swarm of small tremors, or foreshocks, that were first sensed on February 1, and continued with increasing frequency and magnitude over the next three days, reaching a peak on February 4.

At 10:30 that morning, the Liaoning provincial government issued an earthquake warning and began an emergency evacuation of the city of Haicheng. At 7:36 that evening, the Haicheng earthquake struck the area—a

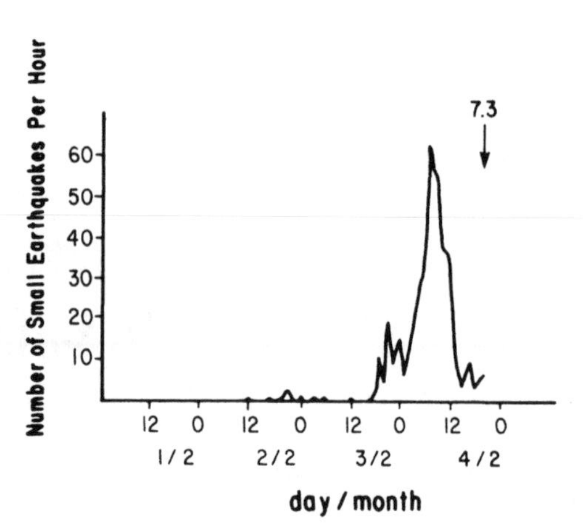

Graph of microseismic activity preceding the Haicheng earthquake of 1975. The arrow marks the time of occurrence of the earthquake of magnitude of 7.3. Note the increase in microseismic activity, which reached a peak just prior to the earthquake. From Yong et al., 1988.

major one measuring 7.3 on the Richter scale. Thanks to the evacuation and other precautions that had been taken, an immeasurable amount of damage and loss of life was averted. It was the first time in history that an earthquake had been forecast with such precision.

It was also the last time, so far.

No earthquake that we know of has changed the course of civilization, as the volcanic eruption at Thera evidently did, but it is safe to say that no geologic phenomenon has taken a greater toll of human lives than earthquakes. They are common, almost daily events at relatively low magnitudes as measured on the Richter scale.

Most of us are familiar with the Richter scale numerology; the newspapers will announce that a quake of magnitude 6.3 was recorded in, say, Peru, or maybe 7.3, which sounds worse. How much worse? The Richter scale, named for Charles Richter (1900–1985), a seismologist who worked at the California Institute of Technology, can be a bit deceptive. It provides a measure of the energy released in an earthquake, determined from seismographic readings taken in local and distant stations, but on an exponential scale rather than a linear one. A magnitude 7.3 earthquake is not just slightly larger than a 6.3 quake; it is *fifty* times larger. In turn, an 8.3 earthquake is fifty times larger than a 7.3 quake, which means that it is *twenty-five hundred* times stronger than a "small" 6.3-magnitude quake.

Even a 6.3-magnitude earthquake is something to be reckoned with. In terms of the energy released, it is the equivalent of a one-megaton nuclear explosion, or about fifty times larger than the Hiroshima bomb. Such quakes are not uncommon (around 100 per year with magnitudes between 6.0 and 6.9), and there are vast numbers of quakes with magnitudes of about 5.3 and less. Most of the global earthquake energy released in any given year, however, is the result of the few earthquakes that register 7.3 and greater.

Obviously, the amount of havoc an earthquake causes depends on where its energy is released. A quake of magnitude 7.3 that occurs at a depth of four miles below the Earth's crust will bring about substantial destruction. On the other hand, a deep focus quake of the same magnitude—one that occurs at, for instance, a depth of 400 miles—may result in little if any damage on the surface. The damage, for the most part, is caused by the motion of the Earth's surface, or ground motion, which will vary considerably depending on the transmissive and dissipative properties of the rocks and geologic formations that lie between the earthquake location, or *focus*, and the surface.

While it has been determined that so far as earthquake (and tornado) damage are concerned, the safest place to live in the United States is near a tiny town called Crossroads in southeastern New Mexico, earthquakes can occur virtually anywhere. Their geographic distribution is generally categorized in the terms of plate tectonics. Thus we have either *interplate* or *intraplate* earthquakes.

Most quakes are of the *interplate* variety, occurring along the boundaries of the Earth's great plates where they grind against each other. The mechanisms of such quakes are fairly well understood in a general way. As one plate moves slowly past its neighbor, enormous strain builds up, not unlike the way in which strain builds up when you try to open a firmly closed jar. Eventually, the strain placed on the lid is enough, and it opens with a pop. Similarly, the strain built up by the plates eventually results in its quick release in the form of earthquake movement, and the plates return to a relatively unstrained state.

The process is more complicated than that, of course, depending on the type of boundary between plates, of which there are three kinds. At mid-ocean ridges, the plates form with the upwelling of magmatic material, and their lateral movement is away from the ridge in opposite directions. This is what is happening at the mid-Atlantic ridge, for example. On the other hand, there are places where two adjoining plates move horizontally relative to each other, usually at different velocities, along what are called *transform* faults (as is the case with the San Andreas fault and others in California). The third boundary type is when one plate is subducted under its neighbor and back into the deep interior of the Earth. Quakes along the mid-ocean ridges

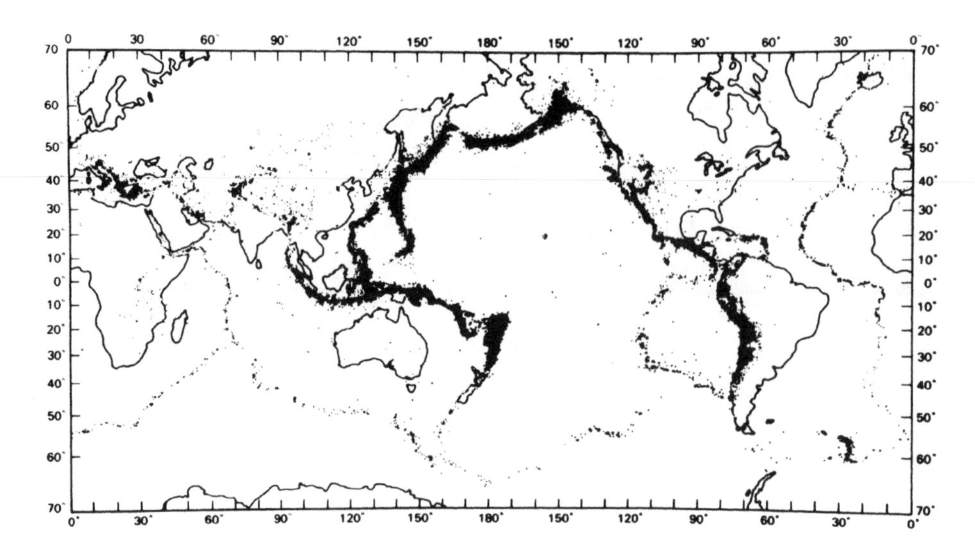

Diagram showing the distribution of earthquake foci. Notice that those in the oceans are concentrated along the mid-ocean ridges and those on the continental borders of the Pacific Ocean along the subduction zones. From Uyeda, 1971.

are relatively few in number and are usually small. Transform-fault quakes can be either small or large in magnitude. The subduction-zone quakes are the most numerous and are often among the largest; their focal zones (that is, where the energy is released) can extend to depths greater than 400 miles. In contrast, the focal zones at the other two boundary types are typically shallow.

The least-understood earthquakes are those that occur within a plate—the intraplate quakes. Though far less common than quakes that occur along the boundaries of plates, the intraplate quakes account for about half of the high-magnitude, shallow-depth earthquakes that rattle into human consciousness and wreak havoc on humankind's works. The Haicheng quake, the only one to have been so successfully predicted that the community could be evacuated, was an intraplate earthquake.

That the first such prediction took place in China seems fitting, since seismology as a science had its origins there with the development in A.D. 132 of the first instrument to record ground motion from earthquakes. Chinese concern with earthquakes is understandable, even at so early a time in history. These violent upheavals are common there and have taken a tremendous toll on the Middle Kingdom's huge population. The State Seismological

Model of the earliest seismograph developed by Zhang Heng in A.D. 132. It consists of a closed bronze urn with a ring of eight dragons holding balls between their jaws. If the earth is tilted by an earth tremor, a pendulum moves inside it and opens the jaws of the dragon facing the source of the tremor. The ball drops into the mouth of the frog sitting below. From Yong et al., 1988.

Bureau of the People's Republic reported that in the thirty-seven years from 1949 to 1976, some 27 million people died and 76 million more were injured as a result of 100 earthquakes. If these figures are correct, the toll is nearly unimaginable. By comparison, the total number of Americans killed in the American Civil War has been estimated at 364,000; those in World War II, at 407,000. The greatest hazard to life in the United States, it is generally agreed, is the automobile, which accounts for 50,000 deaths each year (or in a 37-year period, by way of comparison 1,850,000)—not even a tenth of China's earthquake toll.

In March 1966, two devastating quakes struck China: a 6.8-magnitude quake at Xingtai on March 6, and one of 7.2 at Ningjin on March 22. Premier Zhou Enlai promptly instituted a comprehensive earthquake prediction program and, under this edict and the central control associated with it, China has developed the most extensive earthquake-monitoring system in the world. Though not so advanced in instrumentation as those in the United States and Japan, the Chinese program involves many more people, and more diverse types of potential earthquake precursor signals are tracked.

The science of earthquake prediction is still largely empirical. Scientists measure various properties over time and then examine them *after* a quake to see if any one of them, or several, seems to have been a significant precursor. The hypothesis is that strain will build up over an extended time in a given region and then, during a period of a few months or even years, small cracks (and accompanying tremors) will develop in the underlying rocks. The consequent physical changes in the material properties of the rocks are presumed to extend upward to depths sufficiently shallow that they can be detected by sensitive instruments at the surface.

Most efforts at prediction—in China as elsewhere in the world—concentrate on seismic observations of small earthquake tremors, or foreshocks, which may (or may not) show a recognizable pattern before the main quake. Ground tilting can give some warning over the long term that strain is building up within the earth. As cracks form, changes in groundwater levels will result, as observed in wells. Even the electrical conductivity of the Earth itself changes. There have been reports of sounds and of light, or lightning, just seconds before major earthquakes, but they are poorly documented and the phenomenon is little understood.

The oddest precursor signal, but one that is well documented, is the anomalous behavior of animals, usually occurring over a short time of a few days to a few hours before an earthquake. Most such reports arise in China, where looking for weird behavior among animals is an integral part of the monitoring program (though comparable reports have come from Italy, Japan, and elsewhere in the world). These reports include entries such as this one: "Rats ran away, pandas held their heads and screamed, cattle refused to enter barns or eat grass, dogs became noisy, horses stampeded and neighed, many earthworms came out, fish jumped on shore, mackerel jumped on the sea surface, eels went upstream in many rivers [etc]."

For a good many scientists, such reports are anathema, and they dismiss them out of hand as anecdotal. Among other things, it would be embarrassing to some scientists if the likes of earthworms and eels could pick up precursor signals that sophisticated instruments do not. But the records are arresting and the cause of erratic behavior is not understood. What is one to make of it? Perhaps the best advice is that given by an eminent geophysicist, Harold Jeffreys, some years ago regarding another subject: "If the data speak rot, let the data speak rot." And then set out to understand what phenomena do, in fact, cause such strange behaviors.

After the Haicheng earthquake, it was unclear whether another would soon occur in the general area of northeastern China or whether the strains along the faults were now relaxed. In fact, the Haicheng quake was the last of three great quakes—each above magnitude 7.0 and all local record-breakers in magnitude and intensity—that had struck the region over several years. In June and July, scientists detected a few random precursor anomalies over the nearby area, but they were too inconclusive to issue a warning. There was no swarm of foreshocks of the sort that had been the major tip-off for the Haicheng quake. And, as with responsible earthquake prediction anywhere in populated and industrialized areas, the Chinese seismologists had an additional consideration. As the State Seismological Bureau of China has stated: "Most important of all, because of the political and economic significance of the area, any posting of a warning had to be carefully considered in light of its great social consequences." Who is to take responsibility for evacuating everyone from Los Angeles or Tokyo or Beijing, sending everyone into the countryside and halting all urban activity?

But, in Tangshan, another city in northeastern China, on July 28, 1976, seventeen months after the Chinese had triumphantly predicted the Haicheng quake, lightning flashed across the sky and the Earth rumbled seconds before 3:42 in the morning. And in a matter of seconds, Tangshan, an industrial city of one million people, was reduced to rubble. Almost one-fourth of the city's people perished in the disaster.

Before long, Chinese scientists reviewed their monitoring records. As noted, there had been no swarm of foreshock tremors as in the Haicheng pattern, probably in itself dooming any prediction at Tangshan. But a few precursors had occurred in relative abundance before the earthquake. Groundwater levels had changed and unusual sound and light had been noted, along with electromagnetic anomalies. And animals had acted strangely, indeed in almost exactly the pattern that had shown up at Haicheng. Of all the Tangshan precursors, this is the most enigmatic.

The huge 7.9 magnitude earthquake that devastated much of Sichuan, the mountainous western province of China, at 2:28 local time on May 12, 2008, was remarkable not just for its size and the extent of the damage done to human life and property. It was remarkable for the speed with which the rest of the world became aware of it, Chinese authorities having recently become much more open to the rest of the world. It may be the first major YouTube earthquake: a student at the University of Sichuan filmed his room

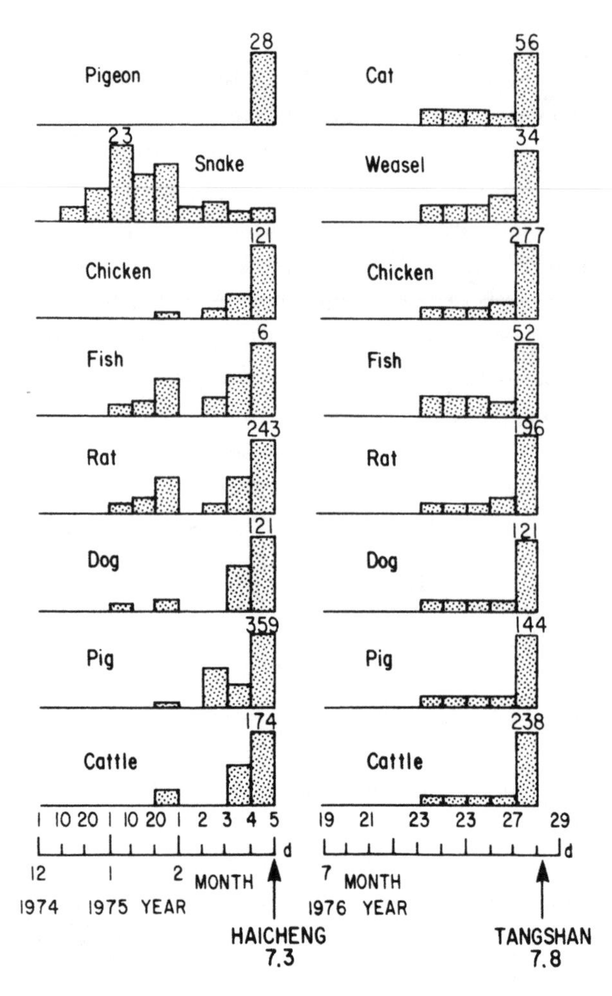

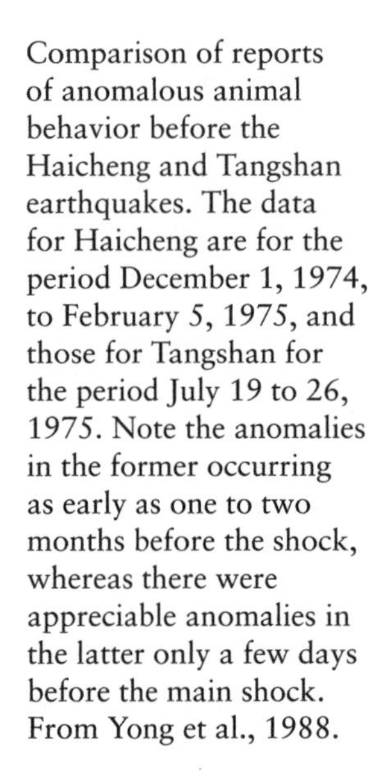

Comparison of reports of anomalous animal behavior before the Haicheng and Tangshan earthquakes. The data for Haicheng are for the period December 1, 1974, to February 5, 1975, and those for Tangshan for the period July 19 to 26, 1975. Note the anomalies in the former occurring as early as one to two months before the shock, whereas there were appreciable anomalies in the latter only a few days before the main shock. From Yong et al., 1988.

shaking, his belongings rattling around the room and his roommate hiding under his desk. The video, available on YouTube within hours, was watched by more than 1,300,000 people around the world.

The epicenter of the quake was twelve miles beneath the town of Yingxiu, some eighty miles from the provincial capital. It was felt as far away as Beijing and Shanghai, where office buildings swayed. Part of the bad news emanating early from China was that many school buildings had collapsed on the students, and it soon became clear that the same had happened to a great many hospitals and factories as well. Apparently the building codes that had been put in place after the Tangshan quake had not been rigorously enforced in some parts of the region as it experienced a big building boom over the past years. Among other things, this made for poor PR for the country soon to host the Summer Olympics.

Another unusual feature of the Sichuan quake was its (and its aftershocks') production of what came to be called "quake lakes." Avalanches caused rivers to be blocked, forming lakes behind the rubble "dams," and these dams all threatened to collapse, causing massive floods to roar downstream, overwhelming innumerable towns along the way. Unprecedented numbers of people were evacuated while authorities tried to disarm some thirty-four quake lakes with explosives and other means. Millions of people were threatened by these lakes.

The number of deaths caused by the quake was given as about 69,000, but this was considered at best a guess. An estimate of property and other damage was $86 billion, making it the world's third most economically expensive natural disaster after Hurricane Katrina in 2005 and the 7 magnitude earthquake that struck the harbor city of Kobe, Japan, in 1995, costing that country $200 billion to put the city back together. Kobe has the misfortune of being located near the juncture of not two but three tectonic plates, each with its own direction.

By comparison with the loss of life in Chinese earthquakes, the toll from America's most famous earthquake was minor: 700. The San Francisco earthquake of 1906, which occurred just before dawn at 5:12 on April 18, was by no means the nation's largest (it is estimated to have been about 7.6 in magnitude) but property losses were enormous: $500 million in current U.S. dollars from the effects of the quake itself and another $3 billion from the fires that ensued.

Among the results were at least two movies about the quake. In the final scene of one of these, Jeanette MacDonald (an opera singer), Clark Gable (a gambler), and Spencer Tracy (a priest) march into the sunset with the rest of the cast, singing "San Francisco," firmly resolved to rebuild the city—an admirable if silly expression of the American pioneer spirit. The great operatic tenor Enrico Caruso was in fact in the city at the time, having arrived to sing *Carmen*. When the quake struck, he ran out of the St. Francis Hotel, one of the few buildings that withstood the earthquake, with a towel wrapped around his neck for protection and cradling a signed portrait of Theodore Roosevelt, evidently for spiritual sustenance. Caruso is reported to have cried out, "Give me Vesuvius!" Upon checking out of the hotel, he vowed never to return to San Francisco, and he never did.

Soon after the earthquake, California set up a State Earthquake Investigation Commission, much the same as the Royal Society had after the Krakatoa eruption, and its report was issued in 1908. Among its findings were a few comments on anomalous animal behavior and earthquake sounds—effects that occurred immediately before the ground motion was felt. In view of the recent Chinese experience, we recite some of them here.

> Horses whinnied or snorted before the shock and stampeded when the latter was felt, some falling owing to the commotion of the ground.
>
> A farmer in the same neighborhood observed his horses moving about, whinnying and snorting, and called to his boy, who was with them, but before the boy could answer he felt the shock.

San Francisco after the earthquake and fire of 1906. United States Geological Survey, Menlo Park, California.

> Several instances were reported where cows stampeded before the shock was felt by the observer.
>
> Dogs generally became alert before the aftershocks, and barked, whined, or ran to cover.

Whether or not domestic animals sensed what was happening at the time, the quake is well understood in plate tectonic terms. It was the result of a break along the San Andreas fault over a distance of 270 miles, centered on San Francisco, at the time a boom city that had been built up rapidly and in a manner of construction that was hardly earthquake resistant. The displacement along the break in the fault was mainly horizontal, in at least one place as much as twenty-one feet.

The San Andreas and its associated fault system extend along much of the length of California from Baja California to Cape Mendocino, north of San Francisco, where it continues out under the sea. It is a plate boundary, a transform-fault system, between the North American plate to the east and the Pacific plate to the west. Relative to the North American plate, the Pacific plate moves toward the northwest at a rate ranging from a fraction of an inch to an inch or so each year. Strain builds up between the plates and, at various locations along the fault system, is released in the form of an earthquake, returning the affected fault to a relatively unstrained state. A quiescent period ensues, followed eventually by more strain and another "adjustment." Throughout history, some portions of the San Andreas fault system have experienced many strain releases in the form of minor tremors, and such regions are considered relatively safe from a large, damaging earthquake. Other areas, like San Francisco, are fated to see quiet periods interrupted by major quakes.

About the time that the Chinese initiated their earthquake-prediction studies (and at a time when the Soviet Union and Japan were enlarging theirs), the United States embarked on its own extensive prediction research program. The direction of other countries' research often has an important impact on decisions about U.S. programs and vice versa. The procedures whereby a new, large-scale scientific study is started in this country follow a set pattern, and the earthquake prediction study was no exception.

First, it is important to have an imprimatur from an august scientific body, more often than not a report from a committee of the National Academy of Sciences, detailing the need for research. Then, the negotiating begins over who will do the work, or lead it: a university, a university consortium,

Fence offset by fault from the San Francisco earthquake of 1906. United States Geological Survey, Denver.

a government agency, or a private corporation? In this case, the decision came down to the United States Geological Survey (USGS) or the Environmental Sciences Service Administration, which subsequently became the National Oceanic and Atmospheric Administration (NOAA). Historically the USGS has been one of the more respected scientific agencies within the government, originally with the primary responsibility of mapping the country. NOAA is a conglomerate of agencies, including the United States Weather Bureau. The final decision for earthquake prediction was made in favor of the Geological Survey.

In the initial stages of such new enterprises, grandiose projections are typically made; in this case, one regularly heard that a prediction scheme was "imminent," though a successful scheme has yet to be developed. Since then, the Geological Survey has made more modest projections. Funding began with about $1 million a year and is now at a level of $20 million. Decisions about funding for this and other such programs are internal, made by the Geological Survey and its parent organization, the Department of Interior. As with research programs cloaked in secrecy in the Department of Defense, it can be difficult for the interested public, or even Congress, to navigate the scientific bureaucracy in order to understand how the levels of funding are determined—and justified. The final result is presented to the public, who can take it or leave it. This is not to say that the system doesn't work. It often works well. But there is little interplay with and review from extenal sources, which can lead to trouble. Sometimes opportunities for innovative research are avoided, or even prohibited. There is an American tendency to believe that if enough money is put into a program, eventually a solution will be found. But this is not always so, nor is it necessarily the best or only approach. In fact, it is often preferable to build a better mouse trap than simply a bigger one.

Like other such studies throughout the world, the U.S. program has sought identifiable precursors to major earthquakes, chiefly in measurements of seismicity and crustal movement and strain. Along the San Andreas fault system, hundreds of highly sensitive instruments feed elaborate data-processing facilities with microseismic information, detecting, counting, and determining the location of every tremor, however slight, and looking for the swarms of small earthquakes that sometimes precede a big one. Lasers on both sides of the faults effectively stare at each other, attuned to detect any relative crustal movement of the Pacific and North American plates. Yet other instruments measure large-scale subsidence and uplift of the Earth's surface. Between the 1960s and 1970s, the surface rose slightly along the San Andreas fault near the city of Palmdale, northeast of Los Angeles. Called the Palmdale Bulge, it is some 130 miles long and is clearly related to the buildup of strain in the underlying rock. Does it presage a major earthquake? It might reasonably, but there has been none so far.

In addition, tiltmeters measure local changes in ground surface, strain meters in drill holes measure vertical strain, and other instruments track water levels in wells, radon gas emissions, and changes in such things as geomagnetism and geoelectricity.

From trenches cut into the land near the fault system, "paleoseismologists" have laid bare the signs of major quakes that have occurred in the past, visible in the way rock strata are displaced. Radiocarbon dating of organic remains in the rock strata provides dates of the ancient major earthquakes, which, they have found, recurred at an average interval of 150 years, with a range from 50 to 300 years.

As the U.S. Earthquake Prediction Program, still in the research phase, is about twenty-five years old, it is not unreasonable to ask how it is doing. A test case is provided by the Loma Prieta earthquake, a 7.1-magnitude quake that struck northern California on October 17, 1989, collapsing the elevated highway in Oakland and a section of the San Francisco Bay Bridge and stunning not only the people of the region but everyone who was watching the World Series between Oakland and San Francisco. The elaborate monitoring system of thousands of instruments busily collecting data all along the San Andreas fault produced no precursors, no foreshock swarms, no anomalous strains or crustal movements. As one scientist put it, "We did not see anything. It is fair to say that it was not encouraging."

But there was an element of serendipity, which often plays an important role in the advancement of science. A few weeks before the earthquake, an independent group of physicists at Stanford University was studying radio noise signals (specifically, low-frequency electromagnetic radiation) using radio receivers located only five miles from the epicenter of the earthquake. About twelve days before the quake, the background radio noise rose, and then, during the three days before the quake, it shot up by a factor of thirty, a level the scientists had never seen in two years of monitoring.

Are these findings happenstance or a real precursor? Why low-frequency electromagnetic radiation would have anything to do with earthquakes is a mystery. But, it should be added, very little is known about the mechanisms involved in the release of earthquake energy, even though it is measured in the range of megaton nuclear bombs.

A few earthquakes have been successfully predicted in the United States, but they have all been small ones, not calling for warnings of any public action. In 1973, scientists at Columbia University began measuring seismic velocities in the Adirondack Mountains of New York. They set off small explosives and watched to see how long it took for seismic waves to reach a nearby receiver. Russian seismologists had reported that anomalous changes in seismic velocities had occurred before some earthquakes, and when the Columbia scientists saw such changes on August 1, they predicted a quake of magnitude 2.5 to 3.0 in the next three days. And it happened on August 3, magnitude 2.6.

Two years later, across the continent in California, a seismograph network was installed to monitor the effects of the additional weight on the

Earth's crust resulting from the construction of a large dam and the filling of the reservoir behind it. (As the reservoir fills, the extra weight of water puts a strain on the Earth's crust that is relieved by land subsidence and, in some cases, more abruptly by earthquakes.) In June, microseismic activity increased significantly and the scientists warned the Department of Water Resources to expect an earthquake. On August 1, a magnitude 5.9 quake shook the area. Then, the following year, in June 1976, the Geological Survey noted an increase in creep along the Calaveras fault east of San Jose, California, and predicted that a quake would occur within a three-month span of time centered on January 1, 1977. The quake, magnitude 3.2, occurred December 6.

Other than these, predictions—especially of major quakes—have not been successful, happily in every instance for the public involved. In 1976, two government scientists (one from the Geological Survey and one from the Bureau of Mines) made a private and unofficial prediction that three major earthquakes of magnitude 8.5, 9.4, and then 9.9 would strike near the coast of Peru in 1981. They based their prediction on unusual criteria. (This is, in fact, one way in which scientific advances can come about: by someone daring to head into the unknown, with the clear risk of being wrong.) Their method involved making gross extrapolations of the effects observed from rock bursts in mines to earthquake magnitude effects. The scientific community not only disapproved of the method, it noted certain problems with the predicters' mathematical equations, and also simply couldn't swallow the idea that two of the earthquakes predicted would be larger than anything observed in historical time. The American scientific community essentially dismissed the predictions out of hand.

But the forecast simply did not go away. Peru, along a plate subduction boundary, is a region subject to many major earthquakes. Some Peruvian scientists kept the pot boiling, seeing in these predictions a chance to obtain added funds to enlarge and improve the country's network of seismographic stations. In the United States, the Peruvian predictions were largely forgotten until November 1980, when, at an informal dinner following a meeting on earthquake prediction, one scientist made an offhand comment about them. Reporters present at the dinner pricked up their ears and the predictions went public. The media gobbled them up and the two scientists originally involved made confirming comments. Once such reports hit the newspapers, they tend to gain a nearly unshakable credibility in the eyes of the public. Not surprisingly, people in Peru were alarmed, and the Peruvian government asked the United States to evaluate the predictions.

A panel of twelve scientists from the academic world and the U.S. Geological Survey, along with expert witnesses, convened in January 1981, a kangaroo court acting as prosecutor, judge, and jury, and for two days cross-examined the two scientists. They concluded, as expected, that there was no scientific basis for the predictions. Still, it was not until the predicted time of the quakes had passed without incident that the Peruvian populace was able to calm down.

The hazards of unwarranted earthquake predictions can be considerable, and understandably so, causing unjustified concern among the public. In 1989, a Reuters report appeared under the headline: "Geologist who forecast quake placed on leave." The story, which appeared shortly after the Loma Prieta quake of October 17, read as follows:

> A geologist for Santa Clara County may lose his job even though he became a media star by predicting the earthquake that hit Northern California last week.
>
> Jim Berkland was put on indefinite administrative leave after a county supervisor said he was not able to separate his "hobby" of predicting quakes and his daily work as a staff geologist.
>
> Berkland predicted in a local newspaper Oct. 13 that an earthquake between 3.5 and 6.0 on the Richter scale would rock Northern California between Oct. 14 and Oct. 21.
>
> He now says that between Nov. 11 and Nov. 18 a quake measuring about 5.5 will hit Northern California.
>
> County supervisor Sally Reed charged that Berkland is "creating fear in the public," by publicizing his predictions.
>
> Berkland says he thinks that totalling up the number of cats and dogs that run away from home in advance of a quake along with measuring the gravitational pull of the sun and moon help forecast the tremors. He claims an accuracy of 78 percent.

There are so many earthquakes of very small magnitude in California that any kind of guessing game will make you right a great deal of the time.

The pull of the Sun and the Moon also figured in the 1990 prediction that made a name (or more appropriate, notoriety) for a New Mexico meteorologist and business consultant, Iben Browning, before his death. He

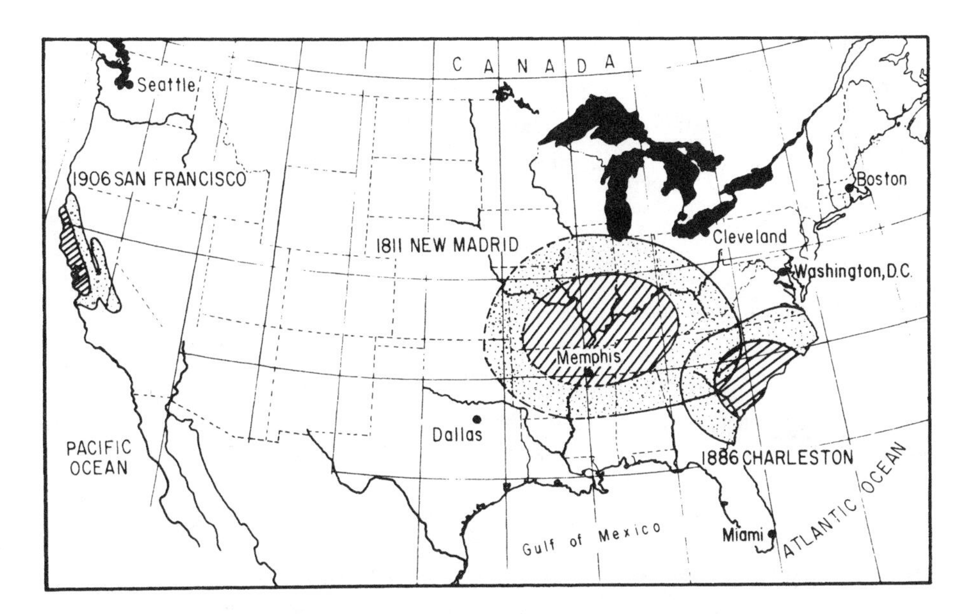

Comparison of isoseismal contours for the New Madrid earthquakes of 1811–1812, the Charleston earthquakes of 1886, and the San Francisco earthquake of 1906. The hachured area represents the zone of most severe ground shaking and damage. The dotted area represents the zone where the damage is less severe but ground shaking is felt by all. From U.S. Geological Survey, professional paper 1240-B, 1981.

predicted a major earthquake on December 3, 1990, in the town of New Madrid, Missouri, causing a local furor of understandable alarm, for it was here that some of the nation's largest earthquakes took place in 1811–12. The other record-setters were the Charleston, South Carolina, quakes of 1886.

The town of New Madrid (locally pronounced *Mad*-rid) is located in the southeastern corner of Missouri on the banks of the Mississippi. When it was founded in 1789 by Colonel George Mason, a patriot of the American Revolution, and a handful of mercenaries, this was a sparsely populated part of the world. The empty lands to the west of the river were under the dominion of Spain, and Mason's notion was to establish a colony under Spanish authority (hence the town's name), which would control traffic along the Mississippi and exact tariffs. But the Spanish ceded the land to the French, who in turn sold it to the United States in the Louisiana Purchase in 1803. New Madrid, its commercial venture foiled, nevertheless prospered as a farming community.

After midnight on December 16, 1811, a church bell pealed in Boston, Massachusetts. It was no pre-Christmas celebration or signal, but the result of an earthquake that rocked New Madrid, 1,100 miles away. The quake occurred at around 2:00 A.M. Another followed on January 23, and yet another on February 7. Their magnitudes have since been estimated at 8.6, 8.4, and 8.7, each one releasing energy equivalent to about 2,500 one-megaton nuclear bombs. Had these quakes occurred a century later, when the entire midsection of the country was far more densely populated, our general consideration of major earthquake hazard regions in the United States might well be different than it is today. (It should be pointed out, however, that earthquakes of this type—major intraplate quakes—tend to recur at intervals of about 600 to 800 years, unlike the 150-year average interval for the San Andreas fault system.)

What is it like when an 8.6 magnitude earthquake hits? The region of extensive damage included parts of Missouri, Illinois, Indiana, Kentucky, Tennessee, Mississippi, and Arkansas. All three quakes were felt throughout the populated areas of the eastern United States. They threw down chimneys in Cincinnati 400 miles away, rattled doors and windows in Washington, D.C., 800 miles away, and, as noted, rang a chuch bell in Boston. Eyewitnesses described the scene at New Madrid as follows:

> The earth was observed to roll in waves a few feet high with visible depressions between. By and by these swells burst throwing up large volumes of water, sand, and coal.
>
> ...Undulations of the earth resembling waves, increasing in elevations as they advanced, and when they attained a certain fearful height the earth would burst.
>
> The shocks were clearly distinguishable into two classes, those in which the motion was horizontal and those in which it was perpendicular. The latter were attended with explosions and the terrible mixture of noises,...but they were by no means as destructive as the former.
>
> Cpt. Sarpy tied up at this [small] island on the evening of the 15th of December, 1811. In looking around they found that a party of river pirates occupied part of the island and were expecting Sarpy with the intention of robbing him. As soon as Sarpy found that out he quietly dropped lower down the river. In the night the earthquake came and next

> morning when the accompanying haziness disappeared the island could no longer be seen. It had been utterly destroyed as well as its pirate inhabitants.

Two waterfalls with drops of about six feet were created in the Mississippi River. Sulfurous or other noxious odors and vapors were noted by virtually every observer in the vicinity. The third (and largest) shock came with a noise so tremendous it was heard as far away as Richmond, Virginia, and the nation's capital. Today, the most conspicuous result of these quakes is still visible: a region of "sunken lands" 160 by 40 miles in southeastern Missouri, western Tennesse, and northeastern Arkansas. There are sloughs, or bottom lands, with depressions of three to five feet; river swamps with depressions of five to ten feet; and lakes of fifteen to twenty feet, including Tennessee's Reelfoot Lake. Here and there in these sunken lands are isolated regions of uplift, circular bulges five to ten miles in diameter.

In Louisville, Kentucky, at the time of the New Madrid quakes, an inventor named Jared Brooks had built a number of pendulums and springs. Curious about earth tremors and vibrations, Brooks had designed the pendulums to detect minor horizontal movements of the Earth, the springs to detect vertical movements. Thanks to his curiosity and his primitive instruments, he was able to record a number of vibrations not generally felt, noting that between December 15 and March 15 there were an astonishing 1,874 aftershocks. The totals varied from a low of 58 for the penultimate week to a high of 292 for the week ending February 23.

A wholly different kind of vibration followed Iben Browning's 1990 prediction of a major quake in the same area. At the time it was forecast, December 3 or 4, schools in Arkansas and Missouri dismissed students, and factories told their employees to stay home. An earth scientist at Washington University in St. Louis later said, "The public was needlessly scared. It was very irresponsible." On the other hand, the media attention accorded to New Madrid by this hare-brained prediction led to an increase in tourism of between 30 and 40 percent. The New Madrid Historical Museum issued T-shirts that read "Visit Historic New Madrid (While It's Still There)" and realized so much money from selling them and other souvenirs that the museum was able to build a $110,000 addition, and began producing New Madrid Fault Line wine.

The series of earthquakes that shook Charleston, South Carolina, in 1886 came in rapid succession, not spread out over almost two months like the New Madrid quakes. The first hit at 9:51 A.M. on August 31 with an estimated magnitude of 8.2. The shock, accompanied by a loud roar and ground motion in the form of waves up to two feet high, was felt in Boston 800 miles away; the upper Mississippi valley, 950 miles off; Cuba, 700 miles to the south; and Bermuda, 950 miles out in the Atlantic. The second shock, of lesser magnitude, followed eight minutes later, and in another ten minutes the third shock occurred, evidently comparable in magnitude to the first. There were two epicentral regions of principal damage—one sixteen miles northwest of the city, the other thirteen miles west of it—each presumably corresponding to one of the two main shocks.

The quakes caused numerous small craters to form, "sand-blows," or "earthquake fountains," as they are sometimes called, which spewed sand ten to twenty feet into the air, covering much of the epicentral region with it. But the damage was not what might be expected from shocks of such magnitude. Relatively few buildings in Charleston were demolished; some suffered only minor damage. How could this be? Probably it was owing to the fact that the ground motion was mostly vertical, rather than horizontal. Horizontal motion causes the greatest damage to structures.

Except that they are all termed earthquakes, the shocks in California and those that occurred in Charleston and New Madrid could hardly be more unalike. California quakes are of the interplate type and are the result of a sudden release of accumulated strain energy. The ground motion in the 1906 earthquake in San Francisco was essentially horizontal. Typically, in California, a single major earthquake is followed by a modest sequence of much smaller aftershocks. In contrast, the Missouri and South Carolina quakes came in sequences of three major shocks of comparable magnitude, accompanied by an extended sequence of aftershocks. In these intraplate quakes, the motion of the ground was essentially vertical.

In both the New Madrid and Charleston earthquakes, loud roars were heard over about as much distance as the shocks themselves were felt. In New Madrid, at least, a sulfurous odor filled the air. We don't know why. We know very little about the mechanisms involved in such intraplate earthquakes, but the two described here illustrate effects that one might expect from a gigantic internal explosion, odd as such a concept may appear.

Far larger in magnitude—and perhaps in its effects on the process of human civilization—was a mammoth seizure of the Earth that took place outside the harbor of Lisbon, Portugal, in the year 1755. Three major shocks struck in rapid succession, at 9:50 A.M. on November 1, and ten minutes later at 10:00 A.M. The third came at noon. The first shock, estimated to

A large craterlet formed by the Charleston earthquake, 1886. United States Geological Survey, Denver.

have been at magnitude 9.0, was the greatest and lasted for six or seven minutes.

At this time, Lisbon was a wealthy capital and the principal city of the Inquisition. It was essentially demolished in the shocks and their aftermath. All of its finest buildings—churches and palaces—were destroyed or severely damaged. Most shops and houses were razed, extensive damage being caused by the resultant fires. In all, 50,000 to 60,000 people were killed. Severe damage occurred throughout the Iberian Peninsula, including Seville, Cordova, and Granada, Spain, and there was damage in North Africa at Fez and Mequinez. Tremors from the Lisbon earthquake were felt in France, Switzerland, Italy, the Netherlands, Germany, and Great Britain and as far north as Fahlun, Sweden, 1,850 miles away—in all, over a territory of two million square miles. And besides the shocks and the fires, a tsunami struck. The sea first retreated in the harbor, then swept back, engulfing much of the city with waves up to fifty feet high. The tsunami reached several places in the North Atlantic: twelve-foot waves broke on the shores of Antigua in the West Indies, 3,540 miles across the ocean.

View of Lisbon following the earthquake of 1755. New York Public Library, New York.

As at New Madrid and Charleston, this was an intraplate earthquake, and for this vicinity it was one of a kind. Nothing of this magnitude has been recorded before or since. The fact that the rapid sequence of shocks occurred on All Saints' Day was by no means lost on the populace; indeed, the toll might have been less had not so many people been at Mass. However great the physical damage it caused, the shock was perhaps greater in the realm of ideas, notably to the religious concept of God as a benevolent deity and to the philosophical notion of humanity and nature being at peace together. Was this divine retribution? Was all lost? This devastating event soon pitted two of the era's greatest thinkers against each other: François Marie Arouet (Voltaire) and Jean Jacques Rousseau.

In the first place, it was clear to many that God in His anger had destroyed Lisbon. With the earthquake occurring during Mass on a solemn church festival, God must have been condemning the irreverent attitude to His services and holy days. Indeed, God had to be saying that the very saints themselves had asked Him to bring His wrath down on Lisbon and its inhabitants. Like Sodom and Gomorrah, the city had to be destroyed. Many of the devout felt that all was lost and that they were powerless in a broken and sinful world. The wrathful God of the Old Testament had replaced the loving Father of the New Testament. Throughout Europe, clergy from all churches echoed the dismay. "There is no divine visitation which is likely to have so general an influence on sinners as an earthquake," said John Wesley, and the Bishop of Chichester dourly commented, "When the Almighty speaks in such tremendous language, he must not speak in vain."

The Jesuit Gabriel Malagrida, saintly, brave, sometimes erratic, and an intimate at one time to the Portuguese royalty, was one of the more outspoken. He wrote:

> Learn, O Lisbon, that the destroyers of our houses, palaces, churches, and convents, the cause of the death of so many people and of the flames that devoured such vast treasures, are your abominable sins, and not comets, stars, vapours and exhalations, and similar natural phenomena. Tragic Lisbon is now a mound of ruins. Would that it were less difficult to think of some method of restoring the place; but it has been abandoned, and the refugees from the city live in despair. As for the dead, what a great harvest of sinful souls such disasters send to Hell! It is scandalous to pretend the earthquake was just a natural event, for if that be true, there is no need to repent and try to avert the wrath of God, and not even the Devil himself could invent a false idea more likely to lead us all to irreparable ruin. Holy people had prophesied the earthquake was coming, yet the city continued in its sinful ways without a care for the future. Now, indeed, the case of Lisbon is desperate. It is necessary to devote all our strength and purpose to the task of repentance. Would to God we could see as much determination and fervour for this necessary exercise as are devoted to the erection of huts and new buildings! Does being billeted in the country outside the city areas put us outside the jurisdiction of God? God undoubtedly wishes to exercise His love and mercy, but be sure that wherever we are, He is watching, scourge in hand.

Such pronouncements by Gabriel Malagrida were anathema to the Marques de Pombal, the man responsible for reestablishing Lisbon, and he asked the Papal Nuncio to banish Malagrida. The Jesuit was jailed and then brought before the Inquisition, that least-civilized aspect of Christianity, who found him guilty as a heretic and put him to death by strangulation in a horrid public display at the torchlight end of an *auto-da-fé*.

Elsewhere, in what had been called Europe's Enlightenment, many were turning away from orthodox Christianity (and its excess baggage, like the

Inquisition) and filling the resultant vacuum with a search for God in Nature. The philosopher of divine Nature was Rousseau; its artists were Turner and Constable; its poets, Coleridge and Wordsworth. Nature was given divine stature, and the more simply and the closer Man lived to Nature, the more virtuous he became. Man and Nature were seen at peace with each other. In "The Tables Turned," Wordsworth could write:

> One impulse from a vernal wood
> May teach you more of man,
> Of moral evil and of good,
> Than all the sages can.
>
> Sweet is the lore which Nature brings;
> Our meddling intellect
> Misshapes the beauteous form of things—
> We murder to dissect.

This sort of beatific philosophizing was too much for the hard-nosed thinker Voltaire. He would have none of it. To Rousseau's philosophy, as expounded in his *Discourse on the Origin of Inequality*, Voltaire replied, "No one has ever used so much intelligence to persuade us to be stupid. After reading your book one feels that one ought to walk on all fours. Unfortunately during the last sixty years I have lost the habit."

With the Lisbon earthquake, Voltaire had a means for scuttling the "*tout est bien*" philosophy. In 1756, he wrote and published his *Poem upon the Lisbon Disaster*, arguing that we have to admit the existence of physical evil in this world. We should henceforth contemplate ruined Lisbon and stop deluding ourselves that Man and Nature are at peace with one another. Instead of the silly cliché "*tout est bien*," the truth was otherwise: "*Le mal est sur la terre.*"

Voltaire went even further in his poem. Everyone, he said, would now have to admit that mankind dare not hope for a safe life in this world under the benevolent protection of a divine providence that could be counted on to reward virtuous behavior. Humans were weak, helpless, ignorant of their destiny, and exposed to fearful dangers. The optimism of the age would have to be replaced by little more than apprehensive hope.

What may the most exalted spirit do?
Nothing. The Book of Fate is closed to view.
Man, self-estranged, is enemy to man,
Knows not his origin, his place or plan,
Is a tormented atom, which at last
Must condescend to be the earth's repast;
Yes, but a thinking particle, whose eyes
Have measured the whole circuit of the skies.
We launch ourselves, like missiles, at the unknown,
Unknowing as we are, even of our own.

Rousseau rejected Voltaire's gloomy conclusions. The optimism Voltaire's poem attacked had, he said, helped him endure the unendurable. Voltaire argued that an omnipotent God could have prevented the Lisbon earthquake; Rousseau countered that God had not done better by mankind because He did not, He could not, exercise control of things through Nature.

Voltaire (1694–1778). British Museum, London.

Jean-Jacques Rousseau (1712–1778). Bibliothèque-Publique et Universitaire Genève, Geneva.

To anyone even minimally influenced by the stunning rise of science since these men had their say, and by the rise of the scientific method of proposing a hypothesis about the nature of reality and then framing tests of those hypotheses, this second-guessing of God and His unfathomable capacities may seem rather quaint. The largely unpredictable violence of earthly seizures appears to such people as the outcome of only partly understood processes that take place without moral judgments about the behavior of the planet's temporary inhabitants. For most such people, evil is a result only of human activity, while planetary processes are neutral, causing what can be thought of loosely as "evil" only to the extent that morally unmotivated events can be imagined to be malevolent. If people living in a city located in a region where earthquakes are almost certain to occur at some point wish to refer to them, when they eventually do happen, as "acts of God," this is their privilege, but such events have little to do with the real world as we now understand it. Bad things can happen to good people.

Nevertheless, the matter raised by Rousseau and Voltaire still haunts us. Is Mother Earth benign? For someone who has suffered through a natural disaster of any kind, it would be hard to float off in Wordsworthian ecstasies and Rousseauvian optimism. On the other hand, this is the only planet we've got. It's a plentiful place, the platform of life—indeed, perhaps the *only* platform for life in the universe—but it is quite capable of taking life abruptly away. Most likely, very few people who suffered through (or perished) in what some call the Boxing Day Tsunami had any idea what Boxing Day meant. It is an old-fashioned, chiefly English notion, wherein Christian churchgoers were encouraged to put small sums of money in a box in the church to be distributed the day after Christmas to the poor. And on the day after Christmas in 2004, the world heard the appalling news that a tremendous tsunami was rolling across the Indian Ocean, claiming untold numbers of lives along the shorelines starting in Indonesia and heading west the 4,791 miles to South Africa. Also called the Indian Ocean Tsunami, it would end up claiming nearly a quarter of a million lives, making it the worst tsunami in recorded history. It was the most notable result of the second worst earthquake to ever be recorded seismographically, a quake of magnitude 9.1 (and maybe slightly more) that lasted a terrifying ten minutes, itself the result of plate subduction off the west coast of Sumatra. Scientists later reported that the quake caused the entire planet to vibrate a half-inch, and it continues to ring like a bell four years later, adding its slowly diminishing waves to the recently understood seismic chaos of the planet, product of its uncountable shocks.

By comparison, the tsunami born of the famous eruption of Krakatoa killed a mere 36,000 people.

The first indication that something big was happening near Sumatra was picked up by Australian scientists at about one, coordinated universal time, an atomic system that gives the time at zero degrees longitude at Greenwich, England. A half-hour later a wave nine meters high struck the remote northwestern Indonesian town of Aceh and roared four kilometers inland, destroying virtually everything in its path, stripping the land naked. In this Muslim town, the survivors would later ascribe the disaster to Allah's anger with them for being lax about their ceremonial duties. (Had they been Christians, they would surely have seen even more ominous signals, given the disaster's proximity to the day of Christ's birth.)

Rippling along at as much as five hundred miles an hour and rising up in lethal grandeur as it reached lower seas and higher ground to send a huge and continuing surge of salt water inland, the tsunami soon struck Sri Lanka, Thailand, other southeast Asian countries, India, and so forth. Its impact on Myanmar, formerly Burma, is unknown, as the military junta in charge of the country would not release any information to the rest of the world, a practice that would not serve them well when, four years later, a cyclone came their way. A handful of deaths occurred almost a day later in South Africa, the result of disturbances of the coastal waters there by the nearly exhausted tsunami.

Probably the worst natural disaster to ever occur, the tsunami brought forth what probably amounts to the greatest single humanitarian aid effort ever. In all, the United Nations led the effort to raise about $7 billion to repair the destroyed infrastructure of the region and to hasten food, water, and medical assistance to the nearly countless needy. This relatively rapid deployment is credited with avoiding the terrible medical problems that can arise in such cases. In this instance, millions of lives were thus saved. One oddball sign of how important this aid effort was is that two American political antagonists, ex-presidents George H. W. Bush and Bill Clinton, spent months together helping UNESCO raise and expend the funds.

Most of us are inclined not to see supernatural reasons for such earthly calamities. The questions they now raise have more to do with the suitability of mankind as a resident of the planet than theology. For example, even this most lethal tsunami might have taken less human life and wrought less destruction—however slightly—except for what is shortsighted coastal land management along most of the northern coastline of the Indian Ocean. For years now, south Asian governments have encouraged the elimination of mangrove forests along the coast and the destruction, even dynamiting, of coral reefs that lie slightly offshore, the latter to make way for shrimp farming, which is in turn one of the worst polluters of coastal waters. Mangroves and reefs would have slowed down the tsunami, however slightly, and lessened the slaughter, but they do clutter up a nice beach, wouldn't you say?

On the other hand, there is simply no way to escape from an incoming tsunami except with the help of an early warning system. Communication between the nations affected was improvisational at best, even among scientific centers devoted to seismic warnings. At the same time, a half-hour's advance warning is enough to get many people off the coast and a few miles inland.

A year after the catastrophe, countless people were still homeless and jobless, particularly the freelance fishing peoples of the coasts whose land was being suborned for development, including beach resorts and other big-time economic projects. The United Nations then held a conference that called for the building of an Indian Ocean Tsunami Warning System that would consist, at the least, of some twenty-five seismic stations to transmit data to twenty-six national tsunami centers. The centers would have the capacity to call for instant evacuations wherever needed. One suggestion (possibly serious) for better alarm systems was that Muslim towns install sirens or other loud noise capacity in the Muslim prayer towers, to be set off whenever a mullah got word from the tsunami center. Much of the system, including the mullahs' sirens, remains to be put in place in 2009, but where it has been built it has evidently already proven its worth. One can only hope.

3

There Have Been Frequent Flooding and Sea-Level Change Events on Earth

Someone once said that there is nothing like a waiting noose to focus the mind. It is possible to argue that the mind is equally focused by waters rising above one's feet of clay. Most of the people in the world still live along watercourses or near seacoasts, as water is a life-sustaining element. But water, impelled by planetary forces and following its own agenda, often rises up and destroys. Floods are so common, in fact, that we tend to lose track of them as a matter of importance in a catalog of natural disasters. Few floods cause great, concentrated losses of life, except the ones that—remote from our lives—carry off tens of thousands of citizens in places like Bangladesh. We are more apt to think of floods as photographs of people making do, sloshing around in a few feet of water in the aftermath of a hurricane or surprisingly heavy rainstorm, their houses full of muck, their tax and other records, perhaps, a soggy mass of useless pulp. The U.S. Weather Bureau tracks hurricanes these days, and local governments, warned well in advance, typically evacuate their citizens. After a century of levee building and other engineering works designed to "tame" rivers, we don't expect our watercourses to rise up from their snug, ordained paths and threaten our lives and livelihoods. But in terms of property damage, floods are by far the greatest natural hazard humans face. In the 1970s, the total cost that society bore from natural disasters of all kinds was $40 billion each year. Of that stunning price exacted by the neutral forces of the Earth, almost half—in fact 40 percent—was a result of floods. In the period from 1844 to 1972 in the United States, there were eighteen floods that caused damage in excess of $500 million. The megafloods alone cost a total of more than $22 billion (see the accompanying table), and figures do not exist to account for all the lesser floods in our history.

We owe our existence to the happy circumstance of being able to interact at appropriate points in an enormous, global series of cycles by which water is transferred from the skies to the Earth and back, but in this vast system of plumbing we have remarkably little control, except—notably—over rivers. Few rivers, for example, in the American West run free, wild, and unpredictable. In Arizona, only the San Pedro River remains undammed, a relatively quiet and narrow tributary that runs north into the Gila and thence into the now utterly prostrate Colorado.

Nature and people compete for space on a flood plain. Flood of March 12, 1963, on the North Fork Kentucky River at Hazard, Kentucky. (Photograph Billy Davis, *Courier-Journal and Louisville Times*.)

Flood Disasters in the United States with Losses in Excess of $500 Million

Year	Stream or Location	Estimated Damage in Millions of Dollars (1973)	Type of Flood
1844	Upper Mississippi	2,030	Rainstorm
1913	Ohio River	900	Rainstorm
1913	Brazos–Colorado, Texas	600	Hurricane
1927	Lower Mississippi	2,500	Rainstorm
1936	Northeastern U.S.	650	Rainstorm
1936	Ohio River Basin	650	Rainstorm, snowmelt
1937	Ohio River Basin	1,700	Rainstorm
1943	Central states	1,250	Rainstorm
1944	Missouri River Basin	720	Rainstorm
1947	Missouri River Basin	640	Rainstorm
1951	Arkansas River Basin	2,500	Rainstorm
1952	Missouri River Basin	550	Rain, snowmelt
1955	Northeastern U.S. from New England to North Carolina	1,550	Hurricane
1955	California, Oregon, Nevada, Idaho	700	Rainstorm
1961	Texas coast	950	Hurricane Carla
1964	Oregon, California, Washington	820	Rainstorm
1965	New Orleans	590	Hurricane, tidal
1972	Entire East Coast of U.S.	3,000	Hurricane Agnes

From Bolt et al., 1975.

But flash floods claim more and more lives and create greater and greater property damage each year in the West, not because of any change in the pattern of storms, but because people build homes and other structures in their occasional paths and (following the dictates of Rousseau) build in the canyons in search of communion with Nature, ever surprised that features of the Earth that were brought about in the first place by the tendency of water to flow downhill, may still, under the right circumstances, be affected by the same process. Under the right circumstances, even our tamest rivers can exact what seems like vengeance.

Of course, from a human viewpoint, some floods are beneficial. Without the annual flooding of the Nile, there would have been no Egyptian civilization, pyramids, Sphinx, Abydos, Luxor, Valley of the Kings, or Abu Simbel. While rivers and river systems were at the heart of most early civilizations (think of the Tigris and Euphrates, and the Yellow River of China), the Nile is one of a kind.

To begin with, mere size sets it apart. It is 4,160 miles long. It arises as the White Nile in the lakes and swamps of central Africa and as the Blue Nile in the highlands of Ethiopia, and the two converge at Khartoum in the Sudan. In other words, it carries the waters of a tropical region to one of temperate climate, creating a highly fertile swath through an otherwise arid land, thanks to its annual floods. And these floods are more reliable than any other river system's in the world. Ancient Egyptian civilization throbbed in response to two metronomes, the daily appearance and disappearance of the Sun, and the annual swelling of the great river.

The reliability of Nile floods is a result of its immense size and the multiplicity of water sources feeding it. If, for example, there were drought in the Ethiopian highlands, the monsoons over central Africa would usually compensate. The Nile of course is now "tamed" by the Aswan High Dam, at best a mixed blessing, but for thousands of years, from the time of the pharaohs through the Roman occupation, its floods permitted Egypt to be the breadbasket of the eastern Mediterranean world, a storehouse for neighboring countries in times of drought and famine. In pre-Aswan times, the Nile would rise to flood stage in southern Egypt by mid-August, spreading out through overflow channels (and then in breeches through low levees) to engulf the adjoining flood basins, bringing fresh soil in the form of silt, along with nutrient enrichment. Within four to six weeks, the enriching floods would have surged to the northernmost realms, and emptied into the Mediterranean. By October, the floodplains in the south of Egypt would again be dry, the northern basins drying up in late November.

In predynastic Egypt, people relied on only this natural irrigation, planting a postflood crop between November and January that would grow well in the waterlogged lands during the generally rainless months of winter. The earliest archaeological evidence of artificial irrigation systems on the Nile dates to about 3100 B.C., just before the unification of upper and lower Egypt and the time of the pharaohs. Natural channels were deepened, some were diverted into dug irrigation ditches, and farmers began to use buckets to haul water up from the Nile to their adjacent fields. The advantages were manifold. The area of annual cropland was thereby increased; water could be held in basins for longer periods after the natural floods; and after the winter crop of cereal grains, vegetables, and flax had been harvested, a second (or summer) crop of fodder and vegetables could be planted.

Early civil engineering, however primitive at first, was advantageous also as a hedge against the specific unpredictability of this most reliable of rivers. Its floods could be counted on, but they varied slightly in timing and they could vary considerably in magnitude. The height of the Nile's flood was a direct measure of the potential harvest that would ensue, and in 1750 B.C. the Egyptians had begun to establish river-gauging stations, often at the sites of temples—what we now call "nileometers." A nileometer is a deep pit or well built on high ground next to the river, with calibrated markings and a stairway leading down for convenient observation of the river levels. So closely were flood height and harvest size correlated that the nileometer readings were used to set the Pharaoh's tax rates for agricultural communities. It is not likely that a more equitable system, if not tax rate, has ever again been devised.

The tumultuous appearance of large amounts of water in one's life is a considerable attention-getter. Indeed, some 500 separate cultures, including those of Greece, China, Peru, and Native America, have a powerful tale of a great flood that has altered the course of their history among the legends and myths by which they described themselves and their origins. In most of them, as with the biblical tale of Noah's Flood, there are but few survivors, perhaps only one family, from which the race thenceforth derives. In the oral history of the Hopi Indians of the American Southwest, there are tales of ancestral villages inundated because of widespread corruption, with only a handful of righteous people surviving to carry on and ultimately generate the Hopi nation of today. So widespread are these tales of flood that they cannot be attributed either to a single global event (which hardly seems likely) or to the dissemination of one tradition alone. Floods are, as noted, widespread bringers of catastrophe, and people living in groups—from small hamletlike collections to small cities—can easily be imagined to have been wiped out or severely disrupted by floods. Ancient Egyptian civilization is notable in not having such a legend: the Nile is so immense that it could buffer any potentially devastating change along the system. Catastrophic floods weren't among the Egyptians' problems.

But floods apparently were a menace for a number of successive groups that arose in that nearby seedbed of civilization and contention, the Middle East. One of the more exciting archaeological studies during the period from 1840 to 1930 was the tracing of the Biblical Flood story back to the Assyrian civilization of 650 B.C., then back to the Sumerian civilization of 2100 B.C., and finally the discovery of evidence for one or more catastrophic

Elephantine Island. Nileometer built into the quay. The marble slabs calibrated to measure the height of the flood date from the nineteenth century. The ancient markings, not visible, are on the facing wall. From Bowman, 1986.

floods at the ruins of the Sumerian cities along the Tigris and Euphrates rivers. The Assyrian and Sumerian accounts are covered in the *Epic of Gilgamesh*, and the similarities between the biblical and Gilgamesh accounts are remarkable.

Let us start with pertinent excerpts from the biblical account of the Noachian flood as given in Genesis (6:11–21; 7:17–23; 8:3–4, 6–12, 18–21):

> Now the earth was corrupt in God's sight and was full of violence. God saw how corrupt the earth had become, for all the people on earth had corrupted their ways. So God said to Noah, "I am going to put an end to all people, for

the earth is filled with violence because of them. I am surely going to destroy both them and the earth. So make yourself an ark of cypress wood; make rooms in it and coat it with pitch inside and out. This is how you are to build it. The ark is to be 450 feet long, 75 feet wide and 45 feet high. Make a roof for it and finish the ark within 18 inches to the top. Put a door in the side of the ark and make lower, middle and upper decks. I am going to bring floodwaters on the earth to destroy all life under the heavens, every creature that has the breath of life in it. Everything on earth will perish. But I will establish my covenant with you, and you will enter the ark—you and your sons and your wife and your sons' wives with you. You are to bring into the ark two of all living creatures, male and female, to keep them alive with you. Two of every kind of bird, of every kind of animal and of every kind of creature that moves along the ground will come to you to be kept alive. You are to take every kind of food that is to be eaten and store it away as food for you and for them." . . .

For forty days the flood kept coming on the earth, and as the waters increased they lifted the ark high above the earth. The waters rose and increased greatly on the earth, and the ark floated on the surface of the water. They rose greatly on the earth, and all the high mountains under the entire heavens were covered. The waters rose and covered the mountains to a depth of more then twenty feet. Every living thing that moved on the earth perished—birds, livestock, wild animals, all the creatures that swarm over the earth, and all mankind. Everything on dry land that had the breath of life in its nostrils died. Every living thing on the face of the earth was wiped out; men and animals and the creatures that move along the ground and the birds of the air were wiped from the earth. Only Noah was left, and those with him in the ark. . . .

At the end of the hundred and fifty days the water had gone down, and on the seventeenth day of the seventh month the ark came to rest on the mountains of Ararat. . . .

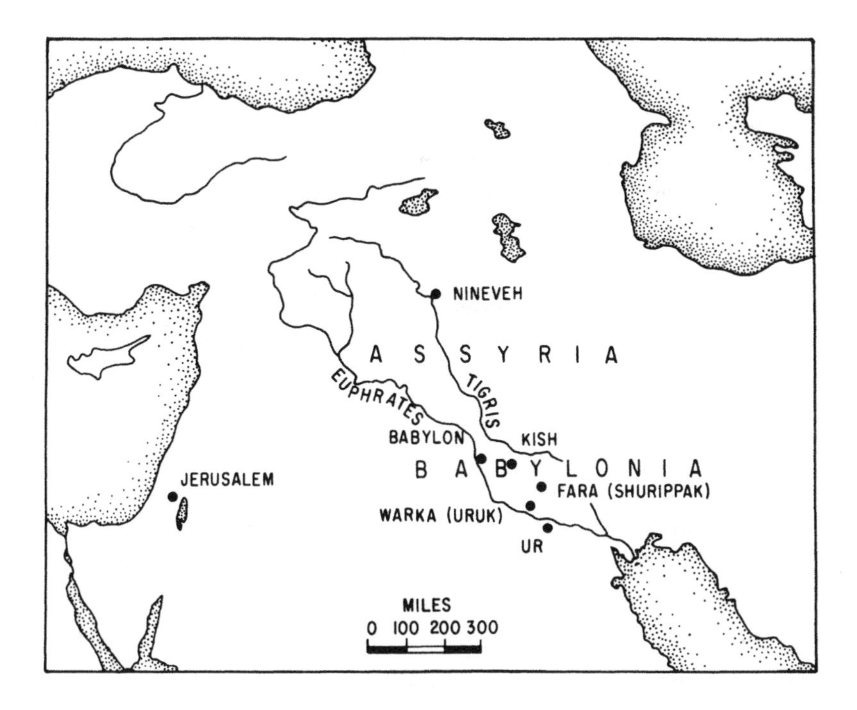

Mesopotamia. Adapted from Kenyon, 1941.

After forty days Noah opened the window he had made in the ark and sent out a raven, and it kept flying back and forth until the water had dried up from the earth. Then he sent out a dove to see if the water had receded from the surface of the ground. But the dove could find no place to set its feet because there was water over all the surface of the earth; so it returned to Noah in the ark. He reached out his hand and took the dove and brought it back to himself in the ark. He waited seven more days and again sent out the dove from the ark. When the dove returned to him in the evening, there in its beak was a freshly picked olive leaf. Then Noah knew that the water had receded from the earth. He waited seven more days and sent the dove out again, but this time it did not return to him. . . .

So Noah came out, together with his sons and his wife and his sons' wives. All the animals and all the creatures

> that move along the ground and all the birds—everything that moves on the earth—came out of the ark, one kind after another.
>
> Then Noah built an altar to the Lord and, taking some of the clean animals and clean birds, he sacrificed burnt offerings on it. The Lord smelled the pleasing aroma and said in his heart: "Never again will I curse the ground because of man, even though every inclination of his heart is evil from childhood. And never again will I destroy all living creatures, as I have done."

The rediscovery of the Gilgamesh story begins with the excavations at Nineveh in 1845. There the library of Ashurbanipal, the last great king of Assyria, who ruled from 668 to 627 B.C., was uncovered. The library consisted of 25,000 tablets, among which was included the Assyrian version of the Gilgamesh epic.

Since then the epic has been found in Akkadian, Hittite, and Hurrian translations dating back to between 1600 and 1000 B.C. The oldest version came from the Sumerian civilization, and the earliest extant writings date back to 2100 B.C. (although the story probably existed earlier, in oral form).

It is important to appreciate that the Gilgamesh story is the earliest major recorded work of literature; Gilgamesh, the title character, was the first human hero in literature. Gilgamesh apparently was a real king at Uruk in Sumer who reigned sometime between 2700 and 2500 B.C.

The Gilgamesh epic is universally appealing because it reaffirms similarities in human nature and values across time and space. Before going on to the portion of the epic that has to do with the Flood it is of interest to quote one other small portion of the epic.

In his travels searching for the meaning of life and for immortality Gilgamesh came upon the fishwife Siduri. At the time, Gilgamesh was sorrowful over the loss of his close friend, Enkidu. The following is their conversation as given in the translation by Donna Rosenberg in her book *World Mythology*.

> Gilgamesh concluded, "Because Enkidu has died, I fear my own death! How then can I be silent? How then can I be still? My friend, whom I dearly loved, has returned to clay! In time, must I also lay my head in the heart of the earth,

> where there are no stars and no sun, and sleep the endless sleep? Oh, Fishwife, now that I have seen your face, do not make me see my death, which I dread."
>
> Siduri replied, "Gilgamesh, where are you roaming? The life that you are seeking you will not find. When the heavenly gods created human beings, they kept everlasting life for themselves and gave us death.
>
> "So Gilgamesh, accept your fate," Siduri replied, "Each day, wash your head, bathe your body, and wear your clothes that are sparkling fresh. Fill your stomach with tasty food. Play, sing, dance, and be happy both day and night. Delight in the pleasures that your wife brings you, and cherish the little child who holds your hand. Make every day of your life a feast of rejoicing! This is the task that the gods have set before all human beings. This is the life you should seek, for this is the best life that a mortal can hope to achieve."
>
> "You have given me wise counsel, Fishwife," Gilgamesh replied.

A little bit hedonistic, but not bad advice for today's society, transported to us from 4,000 years ago.

Gilgamesh then went on to seek the advice of Utanapishtim, the sole survivor of the Great Flood. In the account below, as given again from the Rosenberg translation, Utanapishtim tells Gilgamesh the story of the Flood. In the tale, Enlil is the chief god; Adad, god of the skies; Ishtar, goddess of love and war; Ea, god of wisdom; Shamash, god of the sun; Anu, chief god with Enlil; Nintu, mother of the gods; and Ninurta, the warrior god.

> You are familiar with the city of Shurippak, on the banks of the Euphrates River. When both the city itself and the gods within it were already old, the gods decided to bring forth a great flood. Enlil, ruler of all the gods, called them together in assembly.
>
> "The people who live upon the broad earth have become numerous beyond count, and they are too noisy," he complained. "The earth bellows like a herd of wild oxen. The clamor of human beings disturbs my sleep. Therefore, I want Adad to cause heavy rains to pour down upon the

earth, both day and night. I want a great flood to come like a thief upon the earth, steal the food of these people, and destroy their lives."

Ishtar supported Enlil in his wish to destroy all of humanity, and then all the other gods agreed with his plan. However, Ea did not agree in his heart. He had helped human beings to survive upon the earth by creating rich pastures and farmland. He had taught them how to plow the land and how to grow grain. Because he loved them, he devised a clever scheme.

When Ea heard Enlil's plan, he appeared to me in a dream and said, "Stand by the wall of your reed hut, and I will speak with you there. I will reveal a task for you."

I found myself wide awake, with Ea's message clearly etched in my mind. So I went down to the reed hut and stood with my ear to the wall as the god commanded. "Utanapishtim, king of Shurippak," a voice said. "Listen to my words, and consider them carefully! The heavenly gods have decreed that a great rainstorm will cause a mighty flood. This flood will engulf the cult centers and destroy all human beings. Both the kings and the people they rule will come to a disastrous end. By the command of Enlil, the assembly of gods has made this decision."

"Therefore," Ea continued, "I want you to abandon your worldly possessions in order to preserve your life. You must dismantle your house and construct a giant ship, an ark that you should call Preserver of Life."

"Make sure the ship's dimensions are equal in length and width," Ea counseled. "Build it of solid timber so the rays of Shamash will not shine into it. Take care to seal the structure well. Take aboard your wife, your family, your relatives, and the craftspeople of your city. Bring your grain and all of your possessions and goods. Take the seed of all living things, both the beasts of the field and the birds of the heavens, aboard the ship. Later, I will tell you when to board the ship and seal the door."

I replied, "Ea, my lord, I will do as you have ordered. However, I have never built a ship. Draw a design of this ark on the ground for me, so that I can follow your plan. And when the people of Shurippak ask me what I am doing, how shall I respond?"

Ea then replied to me, his servant, "I am drawing the design of the ship upon the ground for you as you have asked. As for the people of Shurippak, tell them, I have learned that Enlil hates me so that I can no longer live in your city, nor can I place my feet anywhere in that god's territory. Therefore, I will go down to the Deep and live with my lord Ea. However, Enlil intends to shower you with abundance. After a stormy evening, you will find the most unusual birds and fish, and your land will be filled with rich harvests."

With the first glow of dawn, I began to construct my giant ship. The people of Shurippak gathered about me with great interest. The little children carried the sealing materials, while the others brought wood and everything else I would need. By the end of the fifth day of hard labor, I had constructed the framework for my ship. The floor space measured an entire acre. The length, width, and height each measured 200 feet.

I divided the height of the ark so that the interior had seven floors, and I divided each level into nine sections. I hammered water plugs into it and stored supplies. I made the craft watertight. Every day I killed cattle and sheep for the people, and feasted the workers with red wine, white wine, and oil as though they were water from the Euphrates. We celebrated each day as if it were a great holiday!

Finally, on the seventh day, I completed my preparations and moved the ship into the water. When two-thirds of the ship had entered the water, I loaded whatever remained that I intended to take with me. This included what silver and gold I possessed and what living things I had. I put aboard my family and relatives. I put aboard all of the craftspeople. I put aboard animals of the field, both wild and tame.

Shamash had given me a time by which I had to be ready to depart. He had said to me, "When Adad causes the heavens to darken with terrible storm clouds, board the ship and seal the entrance."

So I watched the heavens carefully. When they looked awesome with the gloom of an impending storm, I boarded the ship and sealed the entrance with clay. Long before the storm began to rage upon us, we cast off our ship's cables and prepared to let the sea carry us wherever it would.

The people of the land watched, bewildered and quiet, as Adad turned all that had been light into darkness. The powerful south wind blew at his side, uniting the hurricane, the tornado, and the thunderstorm. It blew for a full day, increasing speed as it traveled, and shattered the land like a clay pot.

In order to observe the catastrophe the heavenly gods lifted up their torches so that the land might blaze with light. But the storm wind raged furiously over the land like a battle. It brought forth a flood that buried the mountains and shrouded the people. No person could see another, and the gods looking down from heaven could not find them either. Its attack ravaged the earth, killing all living creatures and crushing whatever else remained.

As the heavenly gods watched the flood waters pour forth upon the land and destroy everything that inhabited the earth, they too became frightened. They took refuge in their highest heaven, the heaven of Anu. There they crouched against the outer wall, trembling with fear like dogs. Nintu, the mother goddess, wept for the people who lived on the earth.

The goddess Ishtar cried out for the victims of the flood like a woman in labor. "All that used to exist upon the earth in days of old has now been turned to clay," she moaned, "and all because I added my voice to Enlil's in the assembly. How could I agree with the order to attack and destroy my people when I myself gave birth to them? Now the bodies of my people fill the sea like fish eggs!"

Humbled by the enormity of their deed, the heavenly gods wept with Ishtar. For seven days and seven nights the stormy south wind raged over the land, blowing the great flood across the face of the earth. Each day and each night, the windstorms tossed my giant ship wildly about upon the tumultuous sea of flood waters. On the eighth day, the flood-bearing south wind retreated, and the flood waters became calm. Radiant Shamash ventured forth again. He spread his sunlight upon the heavens above and the earth below and revealed the devastation.

When my ship had rocked quietly for awhile, I thought that it would be safe to open a hatch and see what had happened. The world was completely still, and the surface of the sea was as level as a flat roof. All humanity except us had returned to clay. I scanned the expanse of the flood waters for a coastline, but without success.

As Shamash brought his rays of light and warmth inside my ship, I bowed my face to the ground before the powers of the universe. They had destroyed the world, but they had saved my ship. I knelt in submission and respect before Shamash, who nourishes human beings with his healing rays. In gratitude for our survival, I sacrificed an ox and a sheep to the heavenly gods. Then I sat and wept, letting my tears course freely down my face.

My ship floated upon the waters for twelve days. When I next opened the hatch and looked outside, far in the distance in each of the fourteen regions a mountain range had emerged from the surrounding waters. In time my ship came to rest, secure and stable, upon the slopes of Mount Nisir.

For the first seven days, Mount Nisir held my ship fast, allowing no motion. On the seventh day, I set a dove free and sent it forth. The dove could find no place to alight and rest, so it returned to the ship. Next I set free a swallow and sent it forth. The swallow could find no place to alight and rest, so it too returned to the ship. Then I set a raven free and sent it forth. The raven could see that the waters had receded, so it circled but did not return to my ship.

Then I set free all living things and offered a sacrifice to the heavenly gods. I set up fourteen cult vessels on top of the mountain. I heaped cane, cedarwood, and myrtle upon their pot stands, and I poured out a libation to the gods. They smelled the sweet aroma and gathered around me like flies. I prostrated myself before Anu and Enlil.

Then Ishtar arrived. She lifted up the necklace of great jewels that her father, Anu, had created to please her and said, "Heavenly gods, as surely as this jeweled necklace hangs upon my neck, I will never forget these days of the great flood. Let all of the gods except Enlil come to the offering. Enlil may not come, for without reason he brought forth the flood that destroyed my people."

When Enlil saw my ship, he became furious with the other gods. "Has some human being escaped?" he cried. "No one was supposed to survive the flood! Who permitted this?"

Ninurta, the warrior god, said to Enlil, "Do not be angry with us. Only Ea knows everything. Only he could have devised such a scheme!"

Ea then said to Enlil, "You are the ruler of the gods and are wise. How could you bring on such a flood without a reason? Hold the sinner responsible for his sin; punish the person who transgresses. But be lenient, so that he does not perish! Instead of causing the flood, it would have been better if you had caused disease to attack human beings and decrease their number! Instead of causing the flood, it would have been better if you had caused famine to conquer the land. That would have weakened human beings and decreased their number!"

"It was not I who revealed the secret of the great gods," Ea said craftily. "Utanapishtim, the most wise, had a dream in which he discovered how to survive your flood. So now, Enlil, think of what to do with him!"

I bowed my face to the earth in fear and submission before Enlil. He took my hand, and together we boarded my ship. Then Enlil took my wife aboard the ship and made her kneel at my side. He placed himself between us and touched our foreheads to bless us.

"Until now," Enlil said, "Utanapishtim and his wife have been human beings. From this time forward, they will live like the heavenly gods. I have brought down for them everlasting breath so that, like the gods, they may continue to live for days without end. Utanapishtim, the king of Shurippak,

> has preserved the seed of humanity and of plant and animal life. He and his wife will live far to the east, where the sun rises, at the mouth of the river in the mountainous land of Dilmun."

The striking similarities between the Gilgamesh and Noachian legends of the Great Flood indicate a common origin, with the story being passed down from the Sumerians to the Assyrians and the Hebrews (with an important change in the Hebrew version to one God rather than several gods). The story may have been relayed by Abraham and his family in their exodus from the Babylonian city of Ur. Both stories begin with the anger of the gods, or God, with humankind and include divine intervention to save the few humans and to give detailed instructions on how to build the ark. The contents of the ark include not only Utanapishtim, or Noah, and family but also the animals and plants. The storm lasts for seven, or forty, days. The ark comes to rest on Mount Nisir, or Ararat. A sequence of doves, swallows, or ravens is sent out until one does not return. There is remorse on the part of the gods, or God, after the flood and a covenant is made with Utanapishtim, or Noah.

Equally important to the lineage of the Flood legend is the fact that there is direct evidence of floods that devastated the Sumerian cities along the lower Tigris and Euphrates rivers. The following evidence adds a requisite credibility to the supposition that the Flood legend was based on an actual occurrence.

At Ur there is a ten-foot flood deposit of sand and silt. Immediately below the flood deposit, the strata contain a characteristic form of pottery that enables comparison with that found at other sites. The pottery is dated to around 3000 B.C. Above the flood deposit there is evidence of human activity being resumed along lines similar to that of the civilization that existed before. No comparable flood deposit has been found in the later history of the site at Ur.

At Warka two flood deposits have been found with a style of pottery similar to that at Ur. At Kish there is evidence of a flood that occurred considerably after 3000 B.C. At Fara (Shurippak) there is a two-foot layer representing a flood that occurred some time after the one at Ur but before the one at Kish.

A far more tumultuous flood than anything inspiring the Gilgamesh legend may well not have been observed at all. It took place in the eastern portion of what is now Washington state, in an area that has come to be called the Channeled Scablands. The first scientific description of the Scablands was by geomorphologist J. Harlen Bretz in 1923; he described an astounding place and gave an astonishing explanation for it. A labyrinthine pattern of channels had been cut into the bedrock. Basins were cut into the outcropping basaltic rock, some of them up to eight miles wide and 200 feet deep. In some locations, it appeared that 100 to 200 feet of soil cover had been removed. Enormous boulders had evidently been entrained in the channels and rock basins, along with huge chunks of basalt scoured from bedrock exposures. Extinct waterfalls by the hundreds dotted the region, along with spillway gaps connecting spillway basins indicative of flows of water 100 to 200 feet deep. It seemed to Bretz that only a catastrophic flood—not normal stream erosion—could explain these bizarre features. He and a colleague, J. T. Pardee, concluded that the flood occurred when an ice dam holding back the waters of glacial lakes in western Montana failed. For as the Laurentide ice sheet, which covered Canada and a portion of the northern United States, melted over the period between 20,000 and 10,000 years ago, some of the meltwater was trapped in glacial lakes at high elevations.

There simply was not enough energy available in the normal course of stream erosion to carve out such features, and the fact of catastrophic flooding seemed confirmed when, in 1942, Pardee discovered ripple marks on the lake bed of a former glacial lake known as Lake Missoula. Most people have seen ripple marks in the sand at beaches or on river bottoms. But these ripples were twenty to thirty feet high, spaced 200 to 300 feet apart—ripples to match the strides of Paul Bunyan. Pardee calculated that they could have been formed only by a flood of water some 2,000 feet deep, roaring along at the tumultuous velocity of ten cubic miles of water per hour.

While the amazing flood may seem plausible now, if hard to imagine, it was met with scorn by most of Bretz's colleagues in geology (similar to the reaction that Alfred Wegener had encountered when he proposed the plate tectonic theory). Floods were, in fact, a sensitive matter in the history of the geologic sciences. For centuries, the biblical account of the Flood held an important place in geology. It was one of the great catastrophic events that had shaped the known features of the world in its brief existence. But in the eighteenth century, evidence began to appear that the Earth was far older than biblical dates would render it, and geologists like James Hutton began to question what was called *catastrophism*, proposing, in its place, the principle of uniformitarianism. This concept states that the present is the key to the past and that, given enough time, the processes now at work could have produced all of the geologic features of the Earth. It was only in the nineteenth century that the last catastrophists were routed. The final nails in catastrophism's coffin were the reports by John Wesley Powell, who not only led the first party down the Colorado River but also explained how the monumentally large Grand Canyon had been formed by the twin processs of gradual uplift of the land and gradual erosion, both working in tandem over eons. The principle of uniformitarianism was sacred when Bretz indirectly challenged it with his talk of a fantastic flood.

He was invited to give the lead presentation at a meeting of the Geological Society of Washington, D.C., in 1927 on the general topic of the Channeled Scablands. The meeting, however, was stacked with geologists who opposed his ideas, and after his presentation, speaker after speaker, many of whom had never been to the Scablands, expressed their objections to this "outrageous hypothesis." In rebuttal, in an article in 1928, Bretz wrote:

> Ideas without precedent are generally looked on with disfavor and men are shocked if their conceptions of an orderly world are challenged. A hypothesis earnestly defended begets emotional reaction which may cloud the protagonist's view, but if such hypotheses outrage prevailing modes of thought, the view of antagonists may also become fogged.
>
> On the other hand, geology is plagued with extravagant ideas which spring from faulty observations and misinterpretation. They are worse than "outrageous hypotheses," for they lead nowhere. The writer's Spokane Flood hypothesis may belong to the latter class, but it cannot be placed there unless errors of observation and direct inference are demonstrated. The writer insists that until then it should not be judged by the principles applicable to valley formation, for the scabland phenomena are the product of river channel mechanics. If this is in error, inherent disharmonies should establish the fact, and without adequate acquaintance with the region, this is the logical field for critics.

This admonition is as applicable to the earth sciences today as it was in the 1920s. Another, more succinct version was expressed by the nineteenth-century naturalist Louis Agassiz: "Study nature, not books."

The "Spokane flood" was surely one of the greatest floods in the continent's existence and one can only imagine what mayhem it would have created if indeed Spokane had been there at the time.

Most serious flooding in historic times on this continent has been associated with the Missouri–Ohio–Mississippi rivers system, but the greatest natural disasters in the United States have been spawned by hurricanes. The second-worst such natural disaster was the first tropical storm big enough to receive a name in 1992: Andrew. By the time it reached Florida it was a Category 5 hurricane. It caused some $26 billion of damage, devastating much of southern Florida and Louisiana

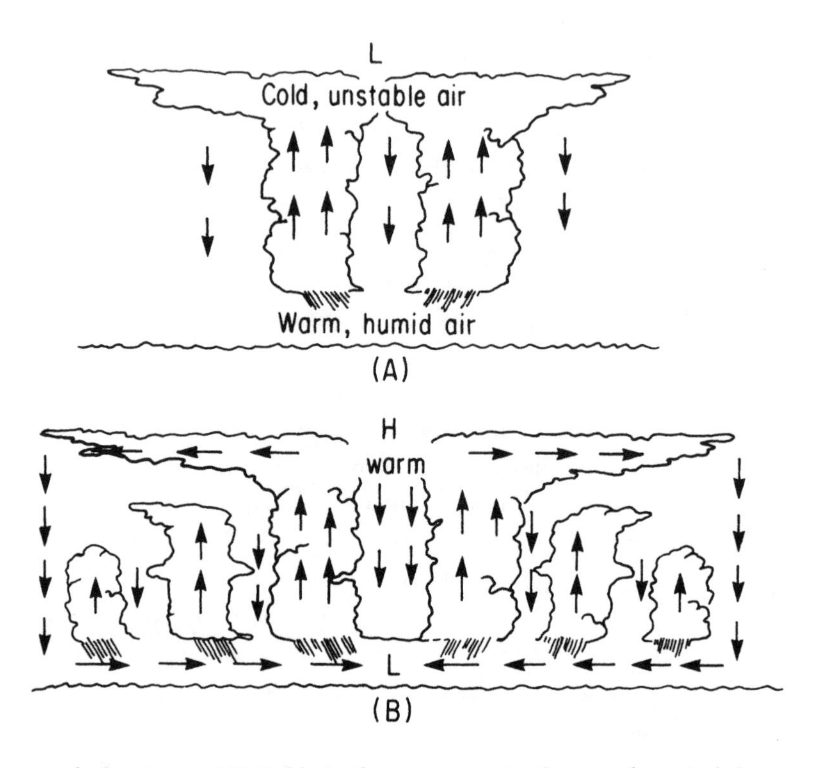

Development of a hurricane. (**A**) Cold air above an organized mass of tropical thunderstorms generates unstable air and large cumulonimbus clouds. (**B**) The release of latent heat warms the upper troposphere, creating an area of high pressure. Upper-level winds move outward away from the "high." This, coupled with the warming of the air layer, causes surface pressures to drop. As air near the surface moves toward the lower pressure, it converges, rises, and fuels more thunderstorms. Soon a chain reaction develops and a hurricane forms. From Ahrens, 1988.

A hurricane (or typhoon) is an especially intense storm that arises in the tropics—between 23½ degrees north and 23½ degrees south of the equator—and packs winds exceeding 74 miles per hour. A hurricane forms over a region where the water temperature is warm and the winds are light. These conditions prevail in the tropics during the summer and early fall, and the hurricane season generally extends from June through November. A hurricane consists essentially of a low-pressure area, or eye, with winds at the ocean surface that, in the Northern Hemisphere, blow counterclockwise around the eye and inward toward its center. Hurricanes blow clockwise in the Southern Hemisphere. The motion is related to the Coriolis force, the centripetal acceleration brought about by the rotation of the Earth; it also causes water in the sink to move counterclockwise down the drain in the Northern Hemisphere, but clockwise in the Southern.

As moist tropical air flows toward the center, it also rises and condenses, creating huge thunderstorms and heavy rainfall. Near the top of the thunderstorms, the air is relatively dry, having lost much of its moisture through rainfall. This drier air moves outward, away from the center, in a clockwise flow. And as the outflow reaches the periphery of the hurricane, it sinks and grows warmer, completing the closed circuit of hurricane circulation.

There is an enormous amount of energy in a hurricane system, much of it latent heat released as water condenses out of the rising air masses. As water changes from a gaseous to a liquid state, energy is released in the form of heat. This process in reverse is more familiar: when water is boiled. It takes energy (heat) to raise the water to the boiling point, but energy is also needed to vaporize it at the boiling point. This added heat requirement is called the latent heat of evaporation. For a mature hurricane, the latent heat released is equivalent to 400 twenty-megaton nuclear bombs; if it could be converted to electricity, it would be ample to supply the needs of the United States for half a year. There is, however, no imaginable way to tame or even divert these prodigious events.

Before Andrew, Hurricane Agnes (another early one) held the record for devastation. It was first spotted over the Yucatan Peninsula of Mexico on June 14, 1972. Five days later, it crossed the Florida panhandle and raged northeastward through the Atlantic states. During its waning stages (when it had been downgraded to a tropical storm), it merged with an extratropical cyclone centered over the Northeast and stagnated over western Pennsylvania for about twenty-four hours from the 23rd to the 24th of June. The entire state was declared a disaster area.

Agnes's winds never rose above minimal hurricane intensity, but it caused disastrous flash floods in an unusually short time over a vast area stretching from Georgia to New York. The storm had been slow to develop and it was exceptionally large in sheer extent, with a diameter of 1,000 miles. It brought with it an unheard-of amount of moisture from the tropics to the East Coast. More than 15 inches of rain fell in a variety of places from Virginia to New York, and 14.8 inches fell in western Pennsylvania in only *24 hours*. Throughout the Middle Atlantic states, record-breaking floods occurred well into July as vastly swollen rivers and river basins—exceeding all previous levels—churned around or over everything in their path to the sea. The Susquehanna River flood was the worst since 1874, and cities such as Harrisburg and Wilkes-Barre were awash for days. Large parts of Richmond, Virginia, were inundated. Nothing like it has happened since, but each fall hurricanes are expected and very carefully watched for along the Gulf and East coasts of the United States. They can transport vast amounts of moisture from the tropical seas and dump it on the land, often accompanied by wind-whipped waves and temporarily high sea levels, that in themselves can do great damage to beaches, dunes, and manmade structures on low-lying coasts.

Since then, what seemed a fairly mild hurricane moving across south Florida became the record-holder in the sweepstakes for the costliest natural disaster to ever strike the United States. It was a Category 1 hurricane called Katrina, and it caused some flooding and even a few deaths with its winds of some 95 miles per hour and a storm surge of no more than five feet on August 28, 2005. But then it left Florida and headed over the Gulf of Mexico, with its much warmer waters.

That it would grow in power and intensity was a certainty; that it would head for the Gulf Coast less of a certainty. But warnings were broadcast, and many Americans watched on television as the hurricane, painted such bright colors on the television screen, indeed headed for land and powered up into a Category 5 storm, the worst there is, with winds up to 175 miles per hour and storm surges expected to be well in excess of eighteen feet, an utter monster if it were to roar in over the beaches of Mississippi and the bayous of Louisiana with virtually nothing higher than a few trees to slow it down.

The Gulf States received a reprieve, if it can be called that. When the storm called Katrina hit land at 7:10 A.M. on August 29, it had shrunk to a Category 3 hurricane—winds up to 125 miles per hour and a storm surge expected up to about 13 feet. (It is often the storm surge that does the most damage.)

It wasn't until Katrina had passed inland beyond New Orleans that the storm surge struck, causing the federal flood protection system there to fail in fifty places. Nearly every levee in the area was breached. Millions of Americans and people around the world watched a major American city drown, as 80 percent of New Orleans wound up under water.

In the days following, the focus of national attention was on those left behind in New Orleans and other nearby municipalities, especially those people who were relocated into the New Orleans Superdome and the city's Convention Center, both of which became what one commentator called "symbols of anguish." On the other hand, an estimated 90 per cent of the residents of the region were evacuated, making it a displacement of Americans that matched those of the Civil War and what some officials called the most successful evacuation of an American city in history.

No one knows exactly how many people left the city or how many decided not to return and have made their homes elsewhere. The death toll was not huge—more than a thousand were accounted for—but the exact number may well never be known, such was the chaos. In economic terms, Katrina was the most destructive natural disaster in United States history, with $60 billion in insured losses and with the real losses more like $125 billion.

To add large-scale insult to catastrophic injury, Hurricane Rita followed Katrina into Louisiana two weeks later. A Category 3 hurricane when it hit the Louisiana and east Texas coasts, it had been the third Category 5 hurricane to careen around the Gulf of Mexico in 2005. Some ten people died as a result of Rita, which caused some $11 billion of damages. Political historians will be assessing and reassessing the political damage from these events well into the foreseeable future.

The greatest flooding of North America, however, was not in the Channeled Scablands or the wake of hurricanes. It is a slower process that is going on even now: the rising level of the sea. On average, over the last 18,000 years, the sea level has risen twenty-six inches every century. Keep in mind that the sea level at a given place can appear to rise because the land is sinking regionally. Or the oceans can actually be rising, which is called a *eustatic* sea-level change, or both. If you drove a stake in the ground at mean tide level on a particular coastline, returned a century later, and found that mean tide level was higher by a few inches, it would be very difficult to tell which phenomenon had caused it. It might be a combination of subsidence (a tectonic change) and eustatic sea-level changes. By averaging observed tide changes globally (assuming that tectonic movements up in one place and down in another average out), scientists have been able to conclude that the sea level has risen eustatically 390 feet over the past 18,000 years. This means, among other things, that a prodigious amount of former real estate now belongs to Neptune.

The cause? Chiefly the Laurentide ice sheet (which created the Scablands), which had a maximum thickness estimated to be 12,000 feet, along with its European counterpart, the Fennoscandia ice sheet. The Fennoscandia covered Norway, Sweden, and Finland and reached 9,000 feet in thickness. An immense amount of water found its way into the oceans as those ice sheets melted. Beginning 18,000 years ago, the rate of sea-level rise was relatively gradual for the next 6,000 years. The sea then rose more rapidly until 7,000 years ago, when it slowed down. Global tide gauge records over the past 100 years show that the present rate of eustatic sea-level rise is around six inches per century.

The 390-foot rise of the oceans has inundated all of the world's continental shelves, which comprise an area equal to 40 percent of the planet's current total land area. Off the East Coast of the United States, the continental shelf extends 50 to 100 miles. Underwater research has shown a series of shorelines where the sea rose and transgressed, then paused (probably

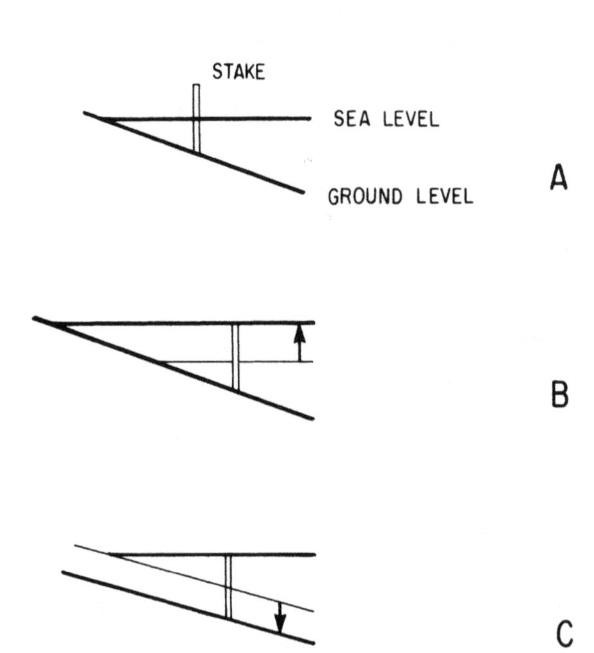

Relative sea-level changes. (**A**) Original sea level and ground configuration. (**B**) Relative sea-level change caused by eustatic sea-level rise. (**C**) The same relative sea-level change caused by tectonic subsidence.

because the land itself rose, creating what is called a "stillstand"—a kind of standoff between rising land and rising sea). Then the sea rose yet again, creating another new shoreline. Along these shorelines, one can see that there has been an overall tilting of the land to the northeast. The New York region has been sinking for 18,000 years at an overall average rate of eight inches per century, while Miami has been uplifted at a rate of eighteen inches a century.

From tide gauge records we can see that today's New York area is subsiding at eight inches a century while the sea is rising at six inches; the net effect is a relative sea-level rise of fourteen inches per century. The same processes tell a different story virtually everywhere in the world, for a vastly complex array of combinations of eustatic sea-level rises and tectonic activity, up or down, or steady.

All of the latest rises in sea level and thievery of the continental shelves has taken place well within the time that humans moved out of the Stone Age and took to agriculture, then to city life. Though it was relatively gradual compared to a human life span, you would nevertheless expect this enormous rise of the waters to be evident in both archaeological remains and in legend. Most of the archaeological evidence is perforce underwater, making it retrievable only by using diving gear and submersibles, expensive vehicles that most archaeologists do not have. But, in the pursuit of legendary remnants, we wind up (possibly) in Eden. This assumes that the Biblical story and the Sumerian legend do represent a land of plenty that is no longer part of the environment. Its location was described in Genesis 2:10–14: "A river watering the garden flowed from Eden; from there it was separated into four headwaters. The name of the first is the Pison; it winds through the entire land of Havilah, where there is gold. The gold of that land is good; aromatic resin and onyx are also there. The name of the second river is the Gihon; it winds through the entire land of Cush. The name of the third river is the Tigris; it runs along the east side of Asshur. And the fourth river is the Euphrates."

We know the Tigris and Euphrates are in Mesopotamia, but what about the Pison and the Gihon? Havilah is located in southwestern Mesopotamia and northern Arabia; in the latter gold was once mined and an aromatic gum called bdellium is still found. Perhaps the Pison is in fact a former river in the area called the Wadi Batin, now dry. The land of Cush has been variously located in Ethiopia and in southeastern Mesopotamia. If it's the latter, then the Gihon may be the Karun, a perennial river flowing out of the highlands

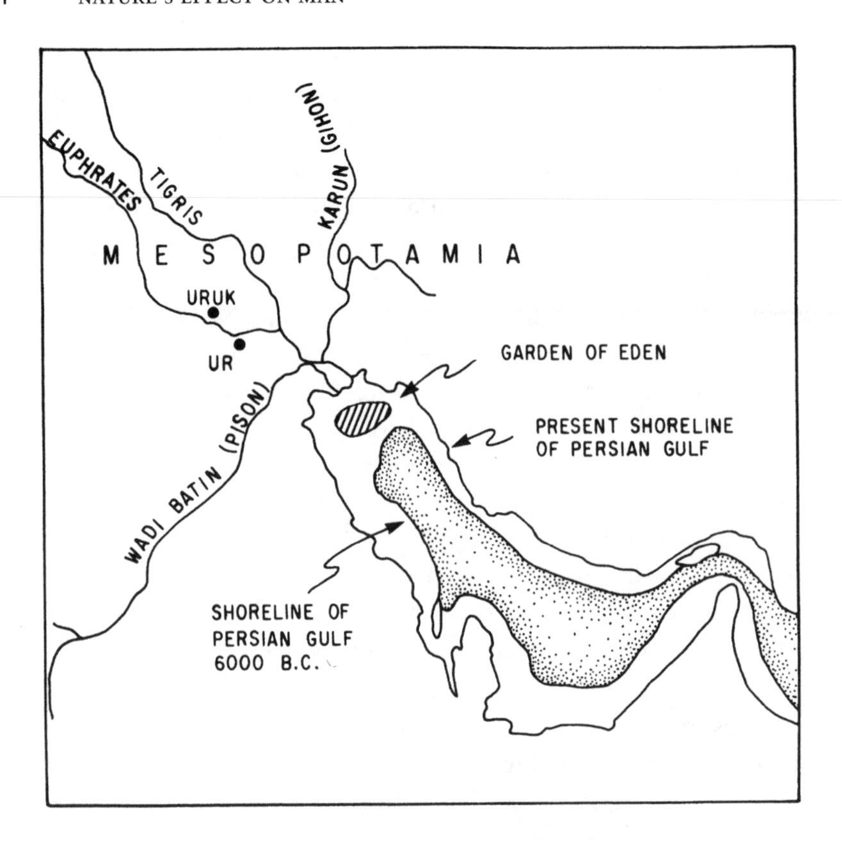

Proposed location of the Garden of Eden. Adapted from Hamblin, 1987.

of Iran that, until it was dammed, contributed most of the sediment that forms the present delta at the head of the Persian Gulf. It is the thesis of Juris Zarins at the University of Missouri that, if there ever was a literal Garden of Eden, it is now underwater (thanks to the glacial rise of the oceans), somewhere in the upper reaches of the Persian Gulf.

When the last glaciers were at their height, and sea level was low, much of the Persian Gulf was dry land. Even as late as 6000 B.C., only about half of the present gulf was underwater. For the next thousand years, paleoclimatic evidence shows that it was a rainy time for this area. If we assume that this legendary primeval land of plenty existed just before or around the time of the rise of the Sumerians (that is, about 5000 B.C.), then we can plausibly say that out in the upper reaches of the Gulf would have lain a wonderfully fertile and productive region that was part of a delta system of four rivers. A place like Eden.

4

... And Occasional Visitors from Outer Space

All it takes is a pair of binoculars to see that the Moon is a pretty inhospitable place, but a quick survey also suggests that it was even more inhospitable in the past. Its surface, and in particular its highlands, is pocked by craters great and small. There are craters within craters, bespeaking devastating events about which astronomers still argue. Clavius, for example, the largest crater visible to us, is 140 miles in diameter and is itself pocked by more than a half-dozen good-sized craters. Some of these craters were probably the result of major volcanic eruptions, but some are surely the result of a titanic battering by the debris that was zipping around in the Solar System when it was forming. Since there never was any water or any atmosphere to speak of on the Moon, probably not much plate tectonic movement, and little erosion, all of these giant pockmarks remain. It would seem unreasonable to imagine that the Earth—the larger of this two-planet system—was not subject to the same disturbances in its youth as the Moon. Of course, the Moon still is struck by debris, but far less so than formerly, now that the Sun's gravity (among other physical forces) has swept up the neighborhood, so to speak, and imposed a more orderly pattern on this tiny region of the universe.

But the regular music of those spheres called planets is still interrupted from time to time by interlopers—notably comets and meteorites. Comets have long been known (people didn't really believe in meteorites until relatively recently) and for most of human history they have been regarded as ill omens. What may have been the comet now known as Halley's appears to have been noted by the Chinese in 1059 B.C. Its arrival was certainly documented in

88 B.C., as well as during each of its subsequent twenty-seven returns in the last two millennia. The comet's vivid appearance in 1066 with a highly visible tail coincided with the Battle of Hastings and, on the Bayeux tapestry that records that historic event in graphic detail, King Harold II of England is shown sitting in obvious discomfort on his throne while alarmed subalterns point to the comet overhead. A bad omen for Harold, perhaps taken otherwise by William the Conqueror. On a later return in 1456 (naturally no one yet knew that it was the same comet), Pope Calixtus III called it "an agent of the Devil."

Not for another three centuries did it come to light that comets—or at least this comet—were a regularly occurring feature of the Solar System. When it appeared in the skies in 1682, the English astronomer Edmund Halley tracked it and determined that it followed an elongated, elliptical orbit around the Sun, an orbit that would take about seventy-five years to complete. He correctly linked its appearance in 1682 with the comets seen in 1607 and 1531, and predicted that it would reappear in 1758. When it did, the scientific community honored him by naming the comet for him. (Other comets are typically named for the first person to see them.)

The return of Halley's comet in 1066 depicted in the famous Bayeux tapestry. Tapisserie Bayeux, Bayeux, France.

Edmund Halley (1656–1742). Bodleian Library, Oxford.

But Halley's honors and certainly his satisfactions were not all posthumous. He was a gentleman-scientist, the equivalent of an endangered species in today's fast-paced and expensive world of science. His research was funded almost entirely from his own and his wife's treasuries. Upon leaving Oxford University with a yen to make a name for himself in astronomy, he set off for two years, with his father's financial assistance, to the island of Saint Helena in the South Atlantic. There he became the first to catalog the stars seen from the Southern Hemisphere. Returning to England in 1678 at the age of twenty-two, he presented his catalog to the Royal Society of London and was promptly elected a Society fellow. He went on to construct his own astronomical observatory and study the planets, stars, and, of course, comets.

Halley was a contemporary and one of the few close friends of the reclusive Isaac Newton, as well as one of the few people of the era who understood the new mathematics Newton was developing, now known as calculus. Halley encouraged Newton to put his work together in a comprehensive statement, which eventually appeared as the *Principia*, arguably the most important scientific treatise ever written. Halley was editor of the volume and, as secretary of the Royal Society, saw to it that it was awarded the Society's imprimatur. And when the Society found it had no funds for its publication, Halley paid for it. Without the *Principia*, Newton would not have come down to us through history as the genius he was. Late in life, he said that he had seen so far because he stood on the shoulders of giants, and one of those certainly was Edmund Halley.

Halley's comet returned again in 1835, on schedule, but relatively little fuss was made over it. The real hoopla took place when it swung into view in 1910, providing an elegant celestial display far more vivid than its most recent appearance in 1985–86. From Accra, in French West Africa (now Ghana), a London correspondent wrote:

> Here everyone has gone mad over it, and we all get up at 4 A.M., and sit and gaze at it till it gets light. It is the most wonderful thing ever seen. The comet itself never rises far above the horizon, but its tail, which stands straight up, is like the rays of a very powerful searchlight—so long, it reaches from the horizon to the very roof of the heavens; and so broad that it occupies roughly one quarter of the arc of the sky; and its light is so powerful that combined with Venus (which is also lovely just now), it has almost the effect of a midnight Sun. The natives are frightened to death of it, and will have that it means an earthquake is coming. . . ."

Gala parties were held in the comet's name in most of the major cities of Europe and America. A comet cocktail was concocted that guaranteed the drinker would see stars. Postcards appeared showing fanciful comets and people's reactions to them. The comet appeared in advertisements for soap, tea, coffee, yeast, light bulbs, fountain pens, chewing tobacco, worm pills, machine tools, soft drinks, beer, and champagne. Politicians, actors, clergy, undertakers, wardens, fire commissioners, and military personnel, as well as scientists, were sought out for opinions. Doomsayers flatly predicted the end of the world. Cyanogen, a deadly gas and minor volatile component of comets, would poison the Earth's atmosphere on its passage through the tail. An Italian flood was laid at the comet's door. A Haitian entrepreneur peddled pills to combat the comet's ill effects.

The great comet of 1843 as seen over Paris. It developed one of the longest tails on record. From Guillemin, 1875.

The notions seem quaint for our time. Yet when forecasts appeared in the early 1970s that a powerfully bright new comet named Kahoutek would show up in the night skies in 1973, the hoopla started all over again. The media came through with various dire predictions backed by some astronomers who should have been more cautious. Many foresaw disaster. The National Aeronautics and Space Administration (NASA) took it as an opportunity to justify the large expenses involved with the Skylab program. Comet T-shirts were popular. Airlines promoted charter flights to observe the comet in case there was bad weather and the comet could not be adequately seen from the ground. The discoverer of the comet, Lubos Kahoutek, a modest and unassuming individual, was persuaded to go out on a special cruise of the luxury liner *QE2* to lecture the onboard comet buffs. But the only untoward event finally associated with Comet Kahoutek was that it was visually somewhat of a fizzle, being nearly invisible.

Alarmists aside, we know a good deal about the nature of comets. For one, they are less important than they look. A comet is chiefly a nucleus that consists of small particles of interstellar dust (about 20 percent of its mass), frozen gases such as methane and ammonia, and water. These irregularly shaped nuclei range from perhaps 100 yards across to a few miles. The rest, what we actually see in the night sky, is largely special effects brought on when the comet nears the Sun. The Sun heats this "dirty iceball," as it has been characterized by the comet professionals. The outer layers of ice melt, releasing some interstellar dust and gas to form the "head" or *coma*. The coma glows in part from its own luminescence and in part from the reflected sunlight—and it can be enormous. The coma of the Great Comet of 1811 was estimated to be more than one million miles across, larger than the Sun itself. A crowd-pleasing comet needs a big, bright coma and, preferably, a tail. The tail is made up of both gas and dust trailing the nucleus. As the London correspondent noted, the tail doesn't follow directly behind the nucleus and coma; it points up, directly away from the Sun. This is because once the comet is near enough to be "lit" by the Sun, it is also within the influence of the solar wind, a continuing stream of ionized particles the Sun gives off in every direction. A major player in the aurora borealis and other earthly phenomena, this stream pushes the more tenuous gas and dust of the tail straight away from the Sun. (It has been noted that the dusty part of the tail may curve somewhat, a compromise between the direction of the comet's motion and the influence of the solar wind.)

Some twenty comets have been observed on two or more periodic appearances and, as Halley did, astronomers have managed to plot their orbits with great accuracy. There seem to be three typical orbits among comets. *Short-period* comets swing around the Sun (the perihelion) and speed on to a farthest point (or aphelion) roughly as distant as the planet Jupiter. *Long-period* comets have their aphelia at Neptune's orbital neighborhood. Then there are comets whose orbits locally appear to be parabolas: they arrive and depart on a curve so great that at first it seems they aren't in an orbit at all. A Dutch astronomer named Jan Hendrik Oort proposed that they *are* in orbits, but vast ones, with the aphelia located in an enormous frozen cloud about one light-year from the Sun. This so-called Oort's cloud remains hypothetical. The theory is that all of the comets we see originated in Oort's cloud, with perturbance by a star or some other cause diverting them into the gravitational arms of the Sun. Some of them, also caught up in turn in the momentary configuration of the gravity of Neptune or Jupiter, are doomed to be exiles in the Solar System.

At least one comet met its end in a plunge too close to the Sun itself. It was labeled XI, not Daedalus, and was photographed by an American satellite in 1979 as it entered the Sun's atmosphere, where a cometary "dust storm" was visible for eleven hours. There is evidence that in the long course of repeatedly orbiting the Sun, a comet's nucleus wastes away until it becomes all tail, in essence a long, tenuous stream of dust along all or most of the former comet's orbit. And as the Earth's atmosphere passes through such an old "tail," some of the dust becomes a meteor shower, such as the annual Perseids shower in August, which gives rise to a beautiful array of shooting stars.

A comet that arrives here and returns to the vicinity of Oort's cloud will take millions of years to return to this part of the Solar System (if it does). At the other extreme is Encke's comet, which metronomically swings through every 3.3 years. It is estimated that there may be 100 billion comets in Oort's cloud, with very few ever likely to leave. It is also estimated that their entire mass is far less than that of the Earth. Indeed, so small are the mass and the density of Halley's comet (or any other) that if the Earth were to pass directly through Halley's head, it would result merely in a noticeable meteor shower, something like the Perseids. The tail's density is equivalent to the vacuum produced in a decent laboratory vacuum apparatus. And of the estimated 2,000 comets in the grip of the Sun's gravity, none comes very close to the Earth: in 1910, Halley's nearest approach was fourteen million miles, a bit farther in 1986.

It would seem that denizens of the Earth have little to fear from these flashy but tenuous "agents of the Devil," though comets cannot be counted out entirely, as we will see later. Closer to home, as it were, and ever so much brawnier, are meteorites.

Meteorites present a *potential* threat to us, however limited. The only person on record ever to be struck by one was a woman in Alabama in 1954, who was hit in the arm when a meteorite plummeted through her roof. They are presumed to come from the asteroid belt, a ring of unassimilated matter left over from the Solar System's origins, out between the orbits of Mars and Jupiter, ranging some 150 million to 500 million miles from the Sun. Asteroids are chunks of rocky or metallic material, or both, and there are some 40,000 of them. The largest known asteroid, Ceres, is 480 miles across. The orbits for about 2,500 are well known; most of them tend to stay put, out beyond Mars, but at least two are in orbits that bring them within less than twice the distance between the Moon and the Earth.

Chunks of material with asteroid-like composition do occasionally make it through the atmosphere and hit the ground, typically the size of a baseball or smaller. Large meteorites are extremely rare; the largest recovered in North America was five feet across. They have been known for millennia and puzzled over for just about as long. Some thought they might be stones carried up by waterspouts and dropped by associated thunderstorms. One of the first scientists to suggest that these anomalous rocks were extraterrestrial in origin was a German physicist named Ernst Chladni, in 1794, and he was met, not surprisingly, with scorn. One colleague said of the theory, "reading it made him feel at first as if he himself had been hit on the head by one of those rocks." Another accused Chladni of being one of those who "deny any world order, and do not realize how much they are to blame for all the evil in the moral world."

Chladni. Wegener. Bretz. A familiar story.

Even Thomas Jefferson, who possessed a fine scientific mind, didn't give much credence to this hypothesis. On December 26, 1807, while Jefferson occupied the White House, a meteorite fall occurred in Weston, Connecticut. Particles from the fall were identified as meteorites by two Yale geologists, who also spoke for their extraterrestrial origin. In perhaps the only example of a president entering directly into a scientific debate, Jefferson commented that he "would rather believe that those two Yankee professors would lie than to believe that stones fell from heaven."

There is an important distinction in terms. A meteor is a piece of interplanetary dust, or something larger, that burns up in the atmosphere with a glow we call a shooting star. When something large enough to make it through the atmosphere hits the ground, it is called a meteorite.

The Earth, it turns out, has also been struck by larger meteorites. In recent years, geologists have found impact crater scars on the Precambrian shield in Canada, a vast tract of ancient rocks, which means they occurred as far back as 1.6 billion years ago. For the most part, the Earth erases such old wounds. But one more recent crater can plainly be seen—indeed, it is a considerable tourist attraction—just south of U.S. Route 40 between Flagstaff and Winslow, Arizona. Called Barringer or Meteor Crater, it is the most definitively studied of all impact craters. Also, it is the most recent and therefore the least eroded, the impact having occurred only some 25,000 years ago. Indeed, the question of whether meteoritic impact was an actual geologic process was largely settled there over a period beginning in the late nineteenth century and stretching well into the twentieth.

To the pioneers of northern Arizona, it was known as Coon Butte, a low-lying circular feature about 4,000 feet across, with a rim that rose as high as 160 feet and an interior depression almost 600 feet deep. Occasionally people snooping around would find fragments of oddly formed iron weighing up to a few pounds on the floor of the depression. Some chunks weighed hundreds of pounds. These reports attracted the attention of Grove Karl Gilbert, one of the founding members of the U.S. Geological Survey, who served as its chief geologist from 1879 to his death in 1918. He was a scholarly, gentle, and even-tempered man.

Intrigued by the reports of what he took to be meteoritic fragments, Gilbert arrived at Coon Butte in 1891 and started with the logical (but as it turned out, incorrect) assumption that if the crater were formed by the impact of a meteor traveling at relatively low velocity, the object would scour out the crater and bury itself beneath the crater floor. If this were true, a magnetic survey across Coon Butte would show a prominent magnetic anomaly. But a survey showed no such anomaly, and Gilbert concluded that the crater was formed in a vast steam explosion of volcanic origin, and that the meteoritic fragments nearby were only incidental.

Then, in a casual conversation in Tucson in 1902, Daniel Moreau Barringer heard about the crater. Barringer was a lawyer by training and a mining engineer of excellent repute by choice. He assumed the crater's origin was meteoritic and, further, that a meteoritic mass of enormous economic value lay waiting to be exploited underneath—an ore body of ten million tons containing nickel, platinum, and iron worth a half-billion dollars. Barringer staked a claim and raised the funds (including some of his personal fortune) for exploratory drilling. He drilled on and off until 1928 and never found the ore body.

But Barringer and a colleague, Benjamin Tilghman, maintained a scientific as well as pecuniary interest in the crater. In the course of their extensive drilling on the site (to depths of 100 feet), they turned up meteorite fragments but found no sign of a volcanic vent. They remained convinced Coon Butte was a meteorite crater, and Barringer was annoyed that the Geological Survey did not acknowledge the validity of his findings or his interpretation. An astronomer tried to assuage him, telling him that he'd have to be "a bit optimistic if he believes that the government will ever admit that it is wrong." Gilbert, though he probably privately accepted Barringer's conclusion, chose to remain silent, and others at the Geological Survey continued to espouse the volcanic origin for several years more.

Aerial view of Meteor Crater, Arizona. John S. Shelton.

Today, it is universally acknowledged that the Coon Butte of old is the result of a meteoritic impact and its new name, Meteor Crater, is a bow to that belief. But what happened to the ore body?

The problem was that both Gilbert and Barringer assumed the meteorite hit at relatively low velocity. The kinetic energy of a moving body is proportional to its mass and also to the square of its velocity. If the velocity is raised by a factor of ten, the mass needed to obtain the same kinetic energy (that is, to produce a depression the size of Meteor Crater) is reduced by a factor of 100. Current thinking is that the object that excavated Coon Butte came in a little faster than an artillery shell and largely vaporized on impact, leaving only scattered fragments, with some forced into the ground below the crater floor. Rather than the temptingly large mass of ten million tons presumed by Barringer, the mass that struck there is thought today to have been only 300,000 tons, an object only 200 feet in diameter. So both men were right in some ways and wrong in others—not an unusual circumstance when scientists delve into the unknown.

Somewhat more ambiguous is the only known *major* impact of an extraterrestrial object in either historical or archaeological time, an event that occurred in the Tunguska River region of Siberia about 400 miles north of Vladivostok on June 30, 1908. That morning, passengers on the Trans-Siberian Railroad were stunned to see a meteor as bright as the Sun race across the sky from south to north and disappear beyond the horizon. Immediately after, they felt a blast of air. It was learned that this blast was felt fifty miles away and flattened forests like matchsticks in a radial pattern twenty miles from the blast center—not that big an event compared to an earthquake, but a spectacular one nonetheless.

The object was assumed to be a meteorite, but subsequent expeditions to this remote site have yet to turn up any meteoritic particles. Something, however, had screamed into the Earth's atmosphere at a shallow angle, something big enough to survive such a long and fiery entry, and apparently exploded several miles above the surface of the ground. Destruction on the ground, it was quickly determined, was not the result of the impact of a solid body but rather the blast wave of an airborne explosion.

What caused the Tunguska event? Nobody knows to this day. Most scientists are content to think it was a small comet, perhaps a chunk of that regular visitor, Encke's comet. It could have been a few tons in mass, or as much as 50,000 tons, which would mean an object about 120 feet in diameter. So it looks like the only major interplanetary havoc-wreaker in modern times—meaning the last 25,000 years—was not a metallic piece of asteroidal artillery, but one of those tenuous iceballs from Oort's cloud.

But maybe not. Since no one seems able to find any physical creations normally resulting from such events, some scientists have looked elsewhere for an explanation. They discount altogether the anecdotal accounts of a light careering in from an angle, and they seek to explain other accounts suggesting that the sky that night lit up, visible from as far away as London, a light that persisted for at least two nights. They suggest that what happened was a supersonic expulsion from under the earth's surface of an immense amount of methane, the shock of which would account for the radial collapse of trees for miles around and which, once ignited, would indeed light up the sky as it burned. While extraterrestrial proponents can't find any physical remains, the methane fans can't find any signs of where the gas might have exited (places called kikberlite pipes) in their proposed earth belch.

Nonetheless, some astronomers supported by NASA have recently offered their services as sentinels to set up a multimillion-dollar system of observatories to watch out for incoming asteroids. The thought of the Earth suffering a major collision with one of the "little planets," with resulting havoc for life itself, has made some people nervous lately. We will return to this controversy after looking into those vagaries of planetary mood known as the climate.

PART TWO

CHANGES IN CLIMATE AND LIFE ON EARTH

5

The Earth's Climate Changes on a Variety of Time Scales

Early in the fifteenth century, around the time of Joan of Arc, the world began to turn cold. All over Europe, the relatively warm and moist climate of what is now called the Medieval Warm Period was replaced by temperatures that averaged 2.5 degrees Fahrenheit lower, a decrease in rainfall to about 90 percent of the previous average, and a greater variability in the seasonal weather from year to year. One hundred and fifty years later, in the mid-sixteenth century, the Flemish artist Pieter Brueghel could paint his countrymen skating on frozen canals; the Thames in London frequently froze over and was closed to all traffic. Called the Little Ice Age, this downturn in world temperatures lasted until the middle of the nineteenth century. It was colder in Valley Forge when Washington overwintered his troops there than it is now, and dwellers in the small southern town that was to become the nation's capital could skate across the Potomac River virtually every winter. By the time the Little Ice Age came to an end, the entire world had been opened up to European investigation and dominance.

The effects of this four-century cold period were not catastrophic, but environments did change, and people were forced to adapt to them, especially those living at high altitudes and in the high latitudes. For example, as the Alpine glaciers of Switzerland and Austria began to advance over the landscape, towns, pasturelands, and mines in the high Alps were abandoned and covered with ice. There was intermittent flooding as ice dams formed glacial lakes, then burst as the glacier continued its advance. A typical account reports: "In 1600, so our ancestors tell us, the big glacier behind

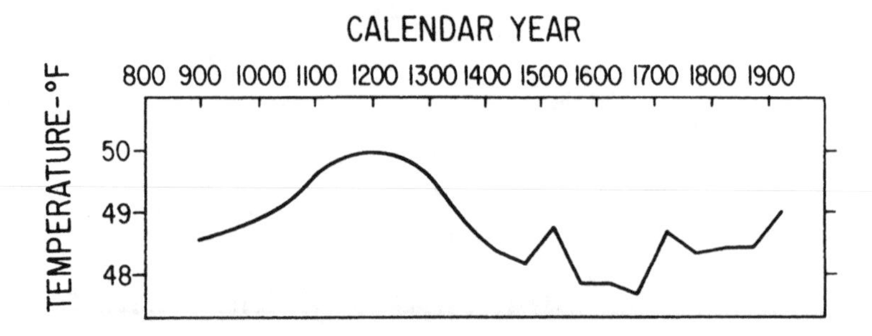

Estimated variation in yearly temperature for central England during the Little Ice Age. From Lamb, 1982.

Rofen after it had come into the valley according to its habit, broke out on the feast of St. James, did great damage to the fields at Ezthal, spoilt the roads and streets and carried away all the bridges. In the parish of Langenfield the water flooded the ground from Rethlstain to Lener Kohlstatt."

The flooding was nothing compared to the magnitude of the glacial lake floods in Washington state that created the Channeled Scablands but certainly was a matter of local concern. As the account suggests, the frontal positions of the glaciers fluctuated, depending on yearly variations in temperature, but around 1850 began a long and continued retreat. Their motions have been carefully documented since about 1600 in drawings, maps, written accounts, and, later, photographs. Their retreat has left behind scoured alpine valleys and deposits of glacial sediments called *moraines*—part of the charm of the region today.

Farther to the north, there had been Norse settlements on the southern coasts of Greenland since Eric the Red in the tenth century, but with the onset of the Little Ice Age, Arctic sea ice began encroaching southward and, for much of each year, Greenland was cut off from communication with Scandinavia. By as early as 1350, the settlements in western Greenland were abandoned, and by 1500 the last settlers on the eastern coast of the island had left; only the native Inuits remained. Iceland, located where the warm waters of the Atlantic gulf stream and the cold waters of the Arctic converge, was only partly affected by the sea ice. Today, as in the Medieval Warm period, the southern coast is ice free, but with the coming of the Little Ice Age, sea ice reached the south several times, cutting off communication with Europe. These were hard times, with economic isolation, loss of farmlands to ice, and the curtailment of ocean fishing.

In Norway itself, the ice advanced relentlessly southward, causing predictable local dislocations. In 1744, one observer wrote:

> Its color is sky blue and it is as hard as the hardest stone could ever be with big crevasses and deep hollow gaps all over and right down to the bottom. Nobody can tell its depth although they have tried to measure it. When at times it pushes forward a great sound is heard, like that of an organ and it pushes in front of it unmeasurable masses of soil, grit and rocks bigger than any house could be, which it then crushes small like sand. In summer there is an awful cold wind blowing off it. The snow which falls on it in winter vanishes in summer but the ice glacier grows bigger and bigger.

As farms were overrun, tax abatement proceedings became a more common occurrence, as in an account of 1735 when a farmer named Guttorm Johanssen Mielvar, of Jostedal Skipreide, appeared in court and asked the people present to attest to the damage the glacier had inflicted on his farm. The people replied "that they could in all God's truth, state and testify that the well known Jostedal glacier, which extends over seven parishes, has grown so much, year after year, that it has carried away not only the greater part of the farmed meadowland but also that the cold given off by the glacier prevents any corn growing or ripening so that the poor man, Guttorm Johanssen who lives there, and his predecessors have each year had to beg for fodder and seedcorn." The neighbors went on to explain that the glacier was now but a stone's throw from the house and that the "farm would be completely carried away within a few years and would never again be habitable." And eight years later, the records explain that the glacier had "crushed the buildings to very small pieces which are still to be seen, and the man who lived there has had to leave his farm in haste with his people and possessions and seek shelter where he could."

In seventeenth century Scotland, fishing, farming, and pasturing all suffered. Some Scots migrated to other countries and professions. During this period, the Scottish mercenary soldier became a familiar figure in the endless clashes of European armies, but the most extensive Scottish migration occurred when James VI of Scotland passed an edict that banished the native Irish from Ulster, and to these more fertile lands the Protestant Scots

The Rhone glacier as seen from the same viewpoint in 1750 (**above**) and in 1950 (**below**). From Lamb, 1982.

migrated in droves. The continuing turmoil and terrorist activity in Northern Ireland is one sorry result of the Little Ice Age that remains with us. This odd climatic glitch is the only period in historical times for which there is evidence of a colder regime worldwide, but the climate has changed often and radically over the long geologic history of the Earth and is continuing to do so, spurred on by our own activities. But before getting to that thorny matter, we will take a look at climate, its change over history, and what we know about the engines of its change.

By climate, we mean the average pattern of weather conditions for a particular region over an extended period. The elements comprised by the weather include air temperature, air pressure, humidity, cloud cover, precipitation, visibility, and wind. Here we are particularly interested in widespread and long-term (that is, climatic) changes in temperature and rainfall.

Climate is a result, in great part, of the heat imbalance of the round Earth. Given the angle at which the Sun's rays strike the Earth, more radiant energy is received near the equator than in the polar regions. Furthermore, much of the radiant energy that does reach the polar regions is promptly reflected back by polar snow and ice. The global tendency is to redress this heat imbalance by redistributing the heat dynamically through the general circulation of the atmosphere. Generally, warmer and less dense air at the equator rises and moves toward the poles, while the cold and more dense polar air sinks and moves toward the equator. This would set up two separate dynamic systems or cells of circulating air except for the fact that the Earth rotates. And this causes each separate equator-to-pole system to break up into three separate circulation cells, rather like sleeves wrapped around the planet: one polar, another at midlatitude (called Ferrel), and one equatorial (called Hadley). The Earth's rotation produces the Coriolis force, which dictates that air in the Northern Hemisphere (which is moving north if it is warm or south if it is cold) is deflected to the right of its natural motion. In the Southern Hemisphere, the effect is the opposite; air is deflected to the left. So, when air at the North Pole moves south, it is deflected to its right, or in a westerly direction, creating winds called the polar easterlies (winds are designated by the direction they blow *from*, rather than to).

In the Hadley cell in the Northern Hemisphere, hot air rises and is replaced by colder air from the north; it too is deflected to the west, creating winds called the northeast trade winds.

Meanwhile, air rising at the southern boundary of the polar cell and air descending at the northern boundary of the equatorial cell produce the Ferrel cell. Here, air at the Earth's surface moves in a northerly direction and is deflected to its right (east), creating what are called the prevailing westerlies. Much of the United States lies under the midlatitude cell with its prevailing westerlies—which is why "weather" tends to come from the west. In the Southern Hemisphere, the north–south directions are flopped, but so is the Coriolis force, the result being polar easterlies, prevailing westerlies, and southeast trade winds.

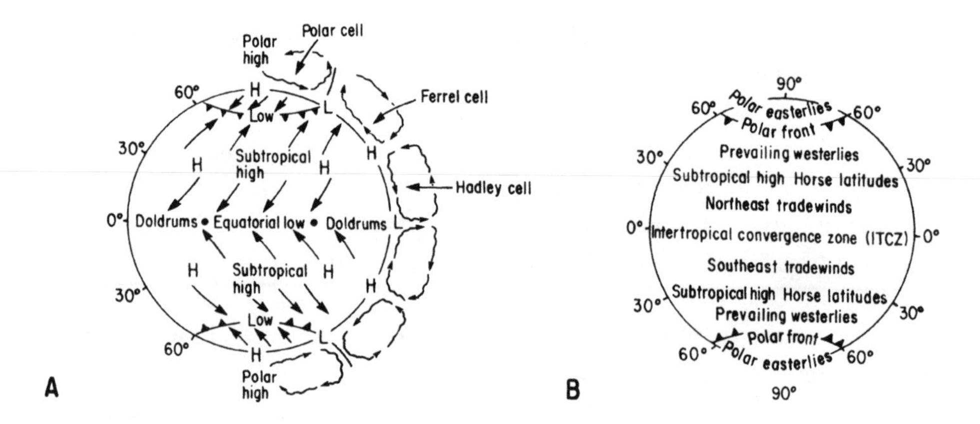

Global atmospheric circulation cells. (**A**) The wind distribution over a water-covered rotating Earth. (**B**) Names of the surface winds for the same system. Adapted from Ahrens, 1988.

In the area of the equator, warm air rises, as we have seen, with some flowing north and some south, leading to the northeast and southeast trade winds. Between these two trade winds, the air at the Earth's surface moves horizontally, but weakly. This area is called the intertropical convergence zone (ITCZ), to sailors of wind-driven vessels, the doldrums. Similarly, the winds are weak where air descends between the equatorial and mid-latitude cells—a zone called the horse latitudes. As in the doldrums, sailing ships were often becalmed here and jettisoned any expendable cargo, presumably including horses. The value of the circular pattern for Atlantic Ocean trade in the days of sailing becomes obvious: ships sailed from Europe with the northeast trades astern, north on the Gulf Stream, and back to Europe with the westerlies. And it is also obvious that with such a dynamic general circulation system, there are changes in the weather and, over time, in climate. Some of these climate changes take place over thousands or even millions of years, and the magnitude of the changes increases with the time scale. But others can be observed over much shorter periods—even tens of years—and some of them take place as a result of the least "promising" parts of the system. For example, the seasonal and longer-term movements of the doldrums are the determining factors defining the changing climatic conditions of the semiarid Sahel region of central Africa, now and for the past two decades so much in the news as a place of intractable drought and famine.

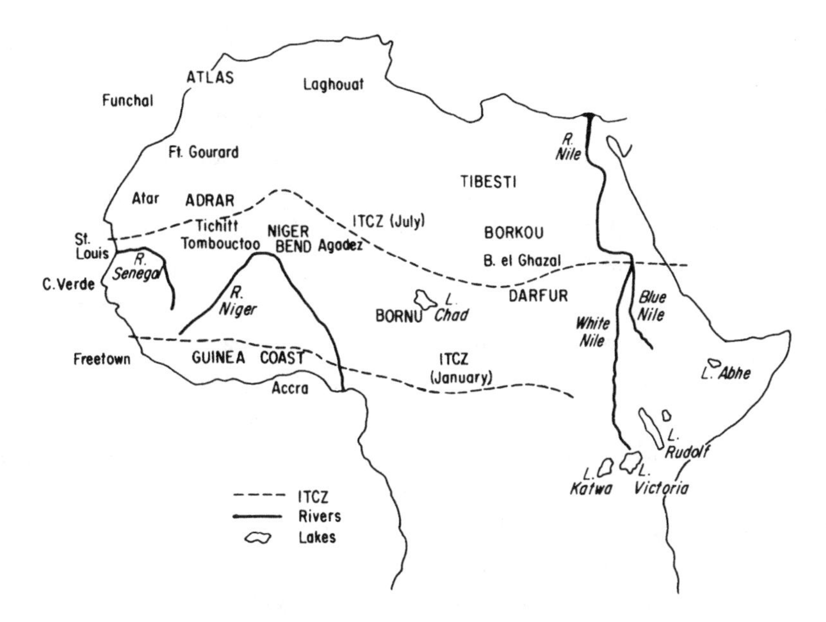

Map of Africa, including the Sahel. The annual excursion of the intertropical convergence zone (ITCZ) from January to July is also shown. Adapted from Williams and Faure, 1980.

The Sahel, which derives its name from an Arabic word for *shore*, is a region that stretches from the eastern to the western shores of Africa (from Ethiopia to Mauritania), but it is also a kind of "shore," or transition zone, between the dry terrain of the Sahara to the north and the well-watered grasslands of central Africa. The present drought in the Sahel became severe in the early 1970s and again in the 1980s, but drought has not been a stranger to the region. In the past, the nomadic populations of the Sahel would head south when drought threatened and return when the rains came again, creating good grazing for their livestock. Today, political boundaries, tribal fiefdoms, civil war, and overpopulation have made migration impossible, and the result has been famine, starvation, and death in heartbreaking numbers. Estimates are difficult, but it is possible that more than one million people have died there since the 1970s from starvation and disease.

Feast or famine in the Sahel is directly related to a fairly delicate seasonal variation in the intertropical convergence zone. North of the ITCZ, the northeast trade winds cross the Sahara, taking dry air with them. To the south, the southeast trade winds carry moist, monsoonal air from the tropical regions of central Africa. Each year for a period of one to four months between June and September, the ITCZ snakes northward, allowing moist air from the south to edge across the Sahel, creating its rainy season. When the ITCZ passes the Sahel on its way south, the rainy season ends, and the dry air of the northeast trades takes over. The degree to which the ITCZ shifts northward and the length of time it remains to the north determines the annual fertility of the Sahel.

Variations in the rainy season can be seen, as at present, in time scales of tens of years. There have been larger variations measured in hundreds of years, corresponding to the Little Ice Age, or in thousands of years, going back to the climax of the last great Ice Age. These variations, large and small, can be read in chronicles and other written records, but also from radiocarbon dating of the successive shorelines of Lake Chad in the central part of the Sahel, shorelines that have risen and fallen to the rhythm of the ITCZ metronome. From the sixteenth to the nineteenth centuries (the Little Ice Age), Lake Chad's water level was twelve feet higher than it is today and, of course, the lake was correspondingly greater in area. Contemporary chronicles tell of long periods of great prosperity for the nomads of the Sahel, interrupted only intermittently by drought.

The water level of Lake Chad was even higher from 4,500 to 6,500 years ago and from 8,000 to 10,000 years ago—times when it was ten times larger than today. Earlier, however, between 12,000 and 20,000 years ago, the area was extremely arid, and the Sahara pushed its dunes south into the Sahel—a time that corresponds to the peak of the last great Ice Age around 18,000 years ago. This long dry spell is readily explained. During that period, much of the northern part of the Northern Hemisphere lay under ice: the Laurentide ice sheet in Canada and the Fennoscandia in Norway, Sweden, and Finland. Meanwhile, the physiography of the Southern Hemisphere was relatively unchanged. The polar cell in the north—with its polar easterlies—extended to the south, a kind of atmospheric domino effect shoving the ITCZ south and suppressing its seasonal urge to move north. The Sahel then became part of the Sahara.

The drought of the 1970s and 1980s in the region appears to be correlated with changes in the equatorial Atlantic and Pacific oceans during the same period, suggesting an oceanic cause. But for the intermediate timescale variations, there is no obvious explanation: only an enigma wrapped in the greater mysteries of the complex system we call global climate.

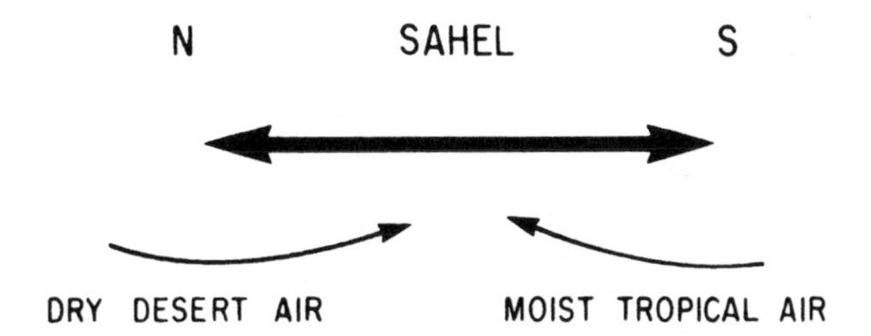

Diagram of seasonal movement of the intertropical convergence zone (ITCZ) in bringing dry desert air to the Sahel when it moves south in the winter and moist tropical air when it moves north in the summer.

The fact is, we know a great deal about the climatic changes that have taken place both over geologic time and over the recent past, but we know very little about their causes. All of the analyses of global temperature over the past century, for example, show a similar pattern—an increase. In the Northern Hemisphere, the average temperature rose about one degree Fahrenheit from 1900 to 1940, then it decreased by half a degree between 1940 and 1970. Since then, it has risen half a degree, returning the Northern Hemisphere to its 1940 state. One degree, half a degree—such changes sound small, but they are within the range we could expect from atmospheric greenhouse effects related to industrial efforts. The recent gain of half a degree since 1970 may well be related to the greenhouse effect, but the earlier rise of one degree cannot be attributed to the greenhouse gases (of which there were markedly less in the atmosphere). And what about the drop between 1940 and 1970? In the 1960s, people were fretting that an ice age was imminent. What is going on? How do we sort out natural global temperature shifts from those caused by human activity?

To confuse matters further, in the Southern Hemisphere the pattern is somewhat different. Rather than the ups and downs, the temperature has risen quite steadily by a total of one degree Fahrenheit over the past 100 years.

Yet another factor in recent global changes is often overlooked: an increase in the variability of seasonal weather during the past thirty years. People tend to emphasize what happened most recently, swearing that the last heat spell was the worst in memory, but careful tabulation and refined recording procedures show a greater variability. The list includes the 1960s,

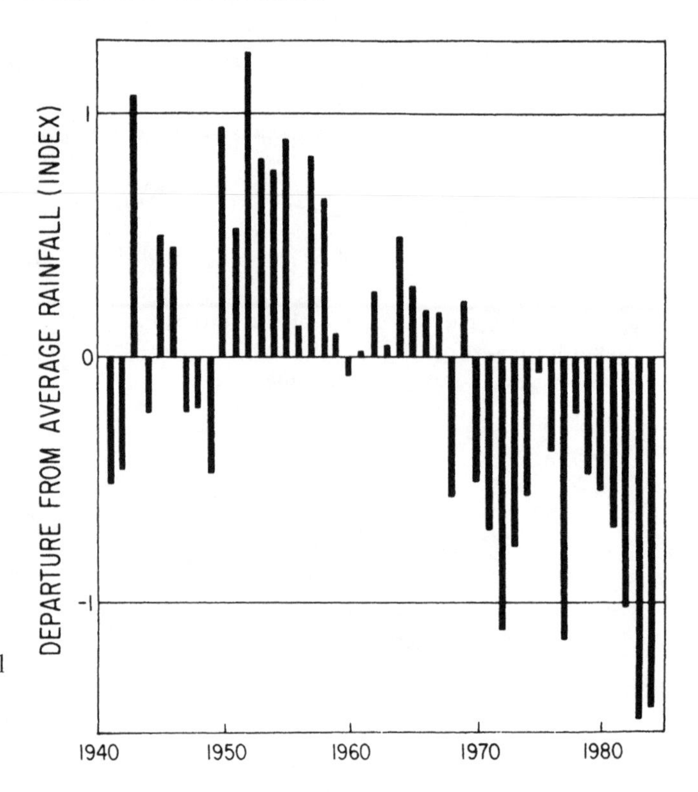

Variations in rainfall for the Sahel, 1940–1985. From Glantz, 1987.

the driest decade in Chile since the 1770s and the 1790s; the winter of 1962–63, the coldest in England since 1740; the winter of 1971–72, coldest in Russia and Turkey in more than two centuries; the heat wave of 1972, greatest on record in Finland and northern Russia; and the winter of 1974–75, the mildest in England since 1834. We could also add the severe drought that struck the midcontinent of North America in 1988.

There are several possible factors controlling such climate changes and we are uncertain about their relative importance. Indeed, there is a philosophical dichotomy between the relative importance of two kinds of factors that might be at work. One set of factors is called *deterministic*, meaning that they are external to the system itself, outside influences that have a changing impact on the climate. The other factors are thought of as inherent in the system: *internal factors*. The philosophical distinction runs deep and has important consequences. For if climate is determined by a range of external factors, we may be able one day to figure out what they

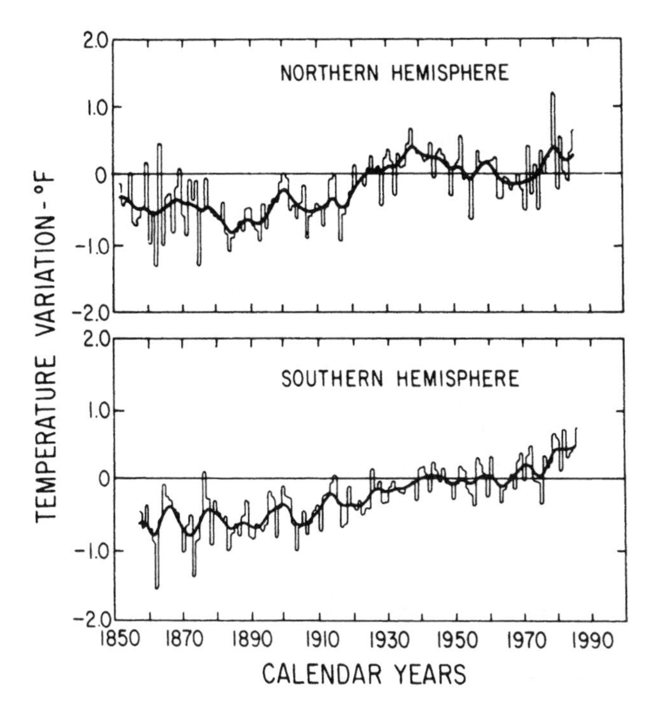

Historical temperature variations for the Northern and Southern hemispheres, 1850–1990. From Michaels, 1990.

are and how they work together, and build models that would permit accurate predictions of climate change. On the other hand, if climate change is related primarily to the internal characteristics of the climate system itself, then accurate predictions are impossible; the best hope is to get a sense of the kinds of changes that can occur at a given time and place. The latter alternative is abhorrent to many scientists, but it remains a distinct possibility.

What are the external forcing functions that can influence climate? The chief ones that have been identified are (1) changes in the energy output of the Sun; (2) astronomical variations in the distance of the Earth from the Sun, and in the angles at which solar radiation reaches various parts of the Earth; (3) changes in the transparency of the atmosphere to either incoming solar radiation or outgoing heat; (4) changes in the circulation systems, the vast currents, of the atmosphere and the oceans; and (5) changes in the earthly surface's absorption or radiation of energy, related to the extent of cloud cover and to the nature of the surface.

Extreme Seasons Reported from 1960 to 1980

Period	Condition
1960–69	Driest decade in central Chile since the 1770s and 1790s
1962–63	Coldest winter in England since 1740
1962–65	Driest four-year period in the eastern United States since records began in 1738
1965–66	Baltic Sea completely ice covered
1968	Arctic sea ice halfway surrounded Iceland for the first time since 1888
1968–73	Severest phase of the prolonged drought in the Sahel zone of Africa, surpassed all recorded twentieth-century experience
1971–72	Coldest winter in more than 200 years of record in parts of Eastern European Russia and Turkey
1972	Greatest heat wave in July in the long records for North Finland and northern Russia
1973–74	Floods beyond all previous recorded experience stretching across the central Australian desert
1974–75	Mildest winter in England since 1834; virtually no ice on the Baltic
1975–76	Great European drought produced the most severe soil moisture deficit that can be established in the London records since 1698
1975–76	Greatest heat waves in the records for Denmark and the Netherlands
1976–77	Severest winter in the temperature records, which begin in 1738, for the eastern United States
1978–79	Severest winter and lowest temperatures recorded in 200 years in parts of northern Europe

From Lamb, 1982.

An obvious candidate for climate change is variation in the Sun's energy output, but regrettably we know little about such changes, even in historical time—much less over geologic time. One solar index (and probably not a very good one) is the number of sun spots, those dark areas seen on the Sun's surface. In historical time, these have been found to vary cyclically with a period of around eleven years. But no one has identified a climatic cycle that occurs on a similar cycle. It is possible that more refined estimates of the Sun's energy output will show a correlation with historical changes of climate, but most scientists are looking elsewhere.

The Earth's orbit does change over long periods of time, thus causing variations in the amount of solar energy that is incident. These variations, known as the Milankovitch cycles, happen at intervals of 22,000, 40,000, and 100,000 years, and they have been invoked as the controlling mechanisms for the advance and retreat of the ice sheets over the past million or so years.

The Earth's atmosphere has undergone changes throughout time, the most dramatic of which were caused by the dust and sulfate aerosol layers injected into the stratosphere by volcanic eruptions, noted in chapter 1. Tambora in 1815, Krakatoa in 1883, and others have produced climatic cooling on regional and even global scales. The more recent eruption of Pinatubo may also have caused an observable climatic cooling.

Certainly, global atmospheric and oceanic currents have changed markedly over geologic time. No Gulf Stream existed before the process of plate tectonics created the Atlantic Ocean. In the shorter term, it is evident that variations in atmospheric and oceanic circulation do occur, maybe on their own, as part of the inherent dynamics of the system itself.

We have little information on how cloud cover has changed over geologic time, but we do know that certain kinds of clouds act as insulators, holding heat near the Earth, while others act as reflectors, sending heat away. We can also be sure that changes in the size of ice sheets, deserts, and vegetated areas (like forests) affect the amount of heat leaving the Earth, which in turn will affect the climate.

Such are the likely external factors that help drive climate change, but assigning relative importance to them, much less actual values, in a global climate model that could be used to make climatic predictions still lies well beyond the capacities of scientists. Beyond that complexity, we have only in recent years come to appreciate that the internal dynamics of such systems may also be an important—if not the most important—factor in understanding climatic changes. This has come to light thanks largely to the rise of a new branch of mathematics known as chaos theory, widely popularized in James Gleick's excellent, accurate, and readable book, *Chaos*.

Atmospheric and oceanic circulation systems are geophysical fluid dynamics systems that function on several different planes of activity simultaneously. Understanding how these activities interact, and with what consequences, poses complicated, almost intractable scientific problems.

Chaos theory, a new mathematics, involves solving *time-dependent, nonlinear, coupled, partial differential equations* to analyze the probable interactions between such physical phenomena as fluid motion, heat conservation, and mass conservation. Although this may seem a daunting prospect, it is a complex process that can be explained in brief form.

The term *chaos* reflects the complexity of the mathematical processes required in pursuing the theory. There are often as many as three equations to be solved simultaneously. Temperature and fluid motion are the dependent terms and appear in each of the three equations; that is, they are coupled. Some of the terms in each equation are products of dependent variables; that is, they are nonlinear. At best, such equations can only be solved analytically for determination of theoretical conditions like equilibrium.

Just how such a system will behave in *time* cannot be arrived at by conventional mathematics. The variety of the factors derived from the fluid dynamics model requires construction of highly sophisticated equations, and their coordinated solution demands powerful computer applications and numerical solution procedures.

Edward Lorenz of the Massachusetts Institute of Technology, when trying to create a mathematical model for atmospheric circulation, found that such elaborate equations can be solved by step-by-step calculations with a computer, giving rise to chaos theory. His calculations were followed by others for other complex fluid dynamic systems, including those by oceanographer Pierre Welander of the University of Washington. Welander used a simple mathematical model consisting of a vertical tube of fluid heated at the bottom and cooled at the top, which is a first-order approximation of an actual convection cell. Since the heated fluid at the bottom tends to rise and the cooled fluid at the top tends to sink, the model can be said to be in a state of dynamic imbalance. Yet, there is a tendency for the system to "seek" stability, or equilibrium. Our intuition tells us that the system might achieve two states of stable equilibrium. Given a slight imbalance in either arm of the loop, the fluid will begin to flow and will presumably keep on flowing in either a clockwise or counterclockwise direction, transferring the heat from the bottom to the top. There is also a possible but unstable equilibrium condition in which there is no motion.

Does the convection system actually work in this way? From the numerical calculations, the answer is no. The clue to this strange behavior lies in the fact that there is an equal chance for the fluid to flow either clockwise or counterclockwise.

The panels, or graphs, on the following page show six ways in which such a sysem can work, in a series of instances where from panel A to panel F there is generally speaking a decrease in heat transfer. The vertical axes of these graphs show flow velocity, while the horizontal axes show time. Where the flow on the graph is above the line labeled 0, the flow is clockwise; in those panels where the flow descends below the line 0, the flow is counter-clockwise.

Panel A gratifies our intuition. Once the flow is started in a clockwise or counterclockwise direction, it increases until it reaches an equilibrium state. Panel B also strokes our intuition; the initial flow may overshoot its equilibrium value, but through a few damped oscillations, it settles back to the equilibrium state. After that, matters are counterintuitive. From panel C on, the numerical calculations of chaos enter the picture.

In panel C, the oscillations do not damp out (as in B) but increase in magnitude until suddenly the flow reverses from clockwise to counterclockwise. Then, the same oscillatory about-face happens and the flow reverses again.

In panel D, yet a different behavior appears. The flow may oscillate one, two, or three times in one direction before it reverses. There is clearly a pattern to this behavior, but there is no way to predict whether a given oscillation will produce a reversal or not. In panel E, the behavior has become even more chaotic, with many variations in the length of time the fluid flows in one direction or another before reversing. Even so, a pattern is discernible.

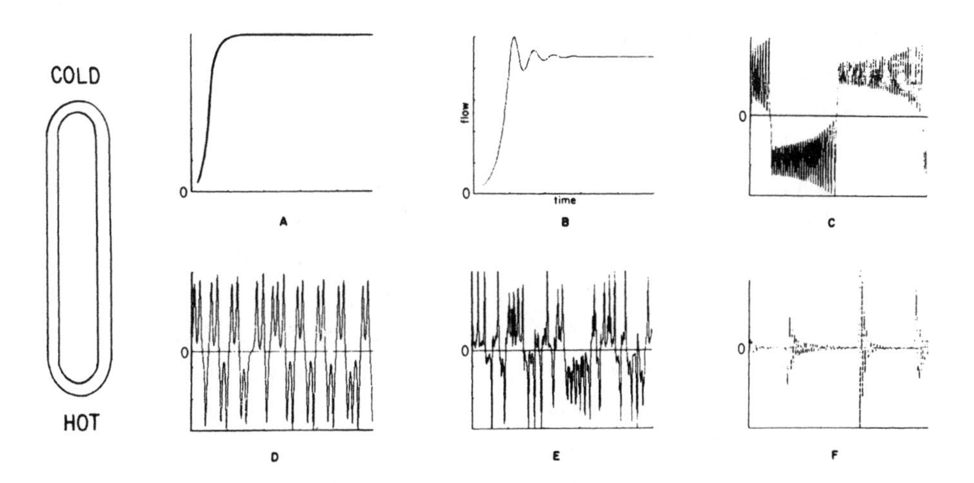

Simple model of a vertical tube of fluid heated at the bottom and cooled at the top (from Welander, 1967) and plots of fluid flow (nondimensional) versus time (nondimensional) for the Welander loop numerical experiment. Progression from panels **A** to **F** corresponds, in a general way, to decreasing heat transfer. From Officer and Drake, 1985.

Finally, in panel F, the unstable equilibrium position of no motion has become, instead, a stable equilibrium with an occasional burp of fluid motion.

That is how complicated things can become in even so grossly simplified a convection cell as a looped tube of fluid. Real circulation systems in the atmosphere or the oceans may or may not show the same kinds of dynamic instabilities, but the example here shows the richness of dynamic variations that *can* exist in nature. It also illustrates that, even with the application of chaos theory, we have a long way to go before understanding climate on its own terms, much less what potentially calamitous changes we are bringing about by our own activities—the much-publicized greenhouse effect. But, that aside, unquestionably the greatest climatic (and geologic) occurrence during the past few hundred thousand years was the ice ages, the most dramatic demonstration of how variable the Earth's climate can be without any "assistance" from humanity. The results of the ice ages lie all around us, defining much of the landscape, and much of the flora and fauna around us, and quite possibly defining, to some degree, our own human nature; it was during the Pleistocene that modern humans evolved.

During the pinnacle of the last Ice Age, around 20,000 years ago, the North American ice sheet covered all of Canada and the northern part of the United States. Across the Atlantic, ice covered all of Scandinavia, the British Isles, and the northern part of mainland Europe. Climates that we now associate with the polar latitudes extended down into the middle latitudes: south of the North American ice sheet was tundra, a treeless marshy plain with short, cool summers. What we now think of as the southern United States was home to boreal forests, the sort of coniferous trees we now see as we approach Canada. In Europe and Asia, the tundra belt was wider than in North America and extended from France to eastern Siberia.

As the ice sheets increased in size and moved southward, they eroded underlying rocks and carried the debris along with them in a vast rearrangement of the Earth's surface. An advancing glacier would gouge U-shaped valleys out of the rock. A tributary glacier entering a larger glacial flow would create a hanging valley; the head of a glacier in a mountainous area would carve out natural amphitheaters called *cirques*. As the ice sheets moved across a crystalline rock surface, they plucked rock debris that acted like great rasps and files, leaving polished rocks, often grooved and striated in the direction of ice sheet movement.

As the ice melted and retreated, it left other features, such as rock basin lakes, and eroded materials in the form of sedimentary deposits. Boulders (glacial erratics) as large as a small house were left behind; tills (unsorted debris) and moraines ("organized" piles of till) were left along the sides or at the end of an ice sheet; and meltwater formed enormous lakes in the gouged glacial topography. The most notable of those that still remain is the Great Lakes. And the source of the water that formed the ice sheets was the world's oceans. The sea level dropped some 390 feet, exposing vast tracts of the continental shelves, which became temporary homes for animals and humans, attested to by mastodon teeth and other remnants dredged up from time to time by fishermen.

That there was more than one ice age has been known for some time from the superposition of till deposits from one ice advance on top of those from a previous one. Alternating ice ages and interglacial periods, including our present interglacial interval, extend back at least one million years, with alternations occurring roughly every 100,000 years. The precise dating of these various ice ages has been made possible only recently through the application of two new techniques, carbon 14 dating and paleomagnetic measurements, but it was not until the last century that scientists would admit that an ice age could have happened at all, much less a regular occurrence of ice ages.

A glacial erratic boulder in Scotland. From Geikie, 1895.

The glacial theory of the Ice Ages originated in Switzerland, not surprisingly, since it is the only country with substantial mountain glaciers in an area in which there has also been a significant number of scientists. The Swiss could see the effects of the present-day Alpine glaciers: huge boulders carried downstream, polished and striated rocks, and morainal sediments. That ice sheets had covered Europe at one time in the ancient past did not seem unreasonable. The great Swiss-born naturalist Louis Agassiz is generally taken as the originator of glacial theory, but the first scientific publications elucidating the role of glaciers were by two lesser-known scientists: Ignace Venetz, a civil engineer, and Jean de Charpentier, a naturalist. Although they should get scientific credit, even they were not the first. In his studies, de Charpentier found anecdotal accounts by Swiss peasants who had realized that such features as glacial erratics at high elevations were the result of past glaciers.

> Traveling through the valley of Hasli and Lungern, I met on the Brunig road a woodcutter from Meiringen. We talked and walked together for a while. As I was examining a large boulder of Grimsel granite, lying next to the path, he said: "There are many stones of that kind around here, but they come from far away, from the Grimsel, because they consist of Geisberger (granite) and the mountains of this vicinity are not made of it."
>
> When I asked him how he thought that these stones had reached their location, he answered without hesitation. "The Grimsel glacier transported and deposited them on both sides of the valley, because that glacier extended in the past as far as the town of Bern, indeed water could not have deposited them at such an elevation above the valley bottom, without filling the lakes."
>
> This good man would never have dreamed that I was carrying in my pocket a manuscript in favor of his hypothesis. He was greatly astonished when he saw how pleased I was by his geological explanation, and when I gave him some money to drink to the memory of the ancient Grimsel glacier and to the preservation of the Brunig boulders.

Through persistent and persuasive advocacy, based on continuing studies in Switzerland, elsewhere in Europe and Great Britain, and in the United States, Agassiz promoted glacial theory. But it took a great deal of promotion and exuberant argumentation with redundant data presented time after time to numerous audiences. For he was up against the biblically derived diluvial theory that attributed the glacial features to the Great Flood of Noah.

Geology has always had an uneasy time with those who prefer a direct interpretation of Genesis, and this was compounded in the 1700s by the fact that many of the Earth scientists, at least in England, were also members of the clergy. An early example of the problem is the date for the Creation that was determined in 1654 by James Ussher, Episcopal archbishop of Armagh and primate for Ireland. From a careful tracking of the biblical begats, Ussher concluded that the Creation took place at precisely 9:00 A.M on October 26, 4004 B.C. This wondrously precise date would probably have passed out of the notice of history except that an unknown scribe included it as a marginal reference note in the King James Version of the Bible. Thereafter, it became doctrine. To assume any other date, especially an earlier one, was heresy. (In fact, if you ignore its exactness, a date of around 6,000 years ago for the Garden of Eden is probably not too bad an estimate, as we saw earlier.)

Early geologists who suspected that more time was involved chose to circumvent the Ussherian dictum by reinterpreting what was meant by each "day" of Creation, but many of them nevertheless clung to the notion of the Great Flood, however untenable it was. For example, they could offer no explanation for how floodwaters had transported boulders the size of a small house and dropped them off 5,000 feet up in the Alps. Eventually the logic and the detailed studies of Agassiz and his converts carried the day.

Now, thanks to paleomagnetism and isotope geochemistry, we know a great deal more about the succession of glaciers that has swelled and shrunk over the past million years. We begin with the fact that Earth is in essence a huge electromagnetic dynamo, given the fluid motion of conducting material in the Earth's core. This fluid motion is controlled in part by the Earth's rotation, and the magnetic pole is roughly the same as the geographic pole of rotation. It might be expected that so dynamic a system would vary over time, and that has indeed been the case; the magnetic North Pole has wandered slightly through even historic time, and its magnitude has decreased by about 5 percent over the past 100 years.

Also, the magnetic poles have changed direction over the eons. At various times, what is now the north magnetic pole, pointing toward geographic north, has been the south magnetic pole. The sequence of reversals has been dated and the paleomagnetic time scale now serves geologists as a reliable backup system for measuring geologic time. This system has proved most fruitful in dating the sequences of events recorded in sediments deep in the ocean. Here, particles have floated down in a relatively quiet environment and, like miniature compasses, the few grains of magnetite and other magnetic materials among the sinking detritus lined up on the bottom in the direction of the Earth's magnetic field. Once covered and cemented to neighboring particles, they were preserved as reliable indicators of the magnetic field at the time they settled there. The observed sequence of magnetic-particle reversals in a given sedimentary core can then be compared with the standard reversal sequence to provide accurate dating of each level in the core. The standard sequence was derived by dating the reversals from the "main" geologic time scale (determined from radioactive age determinations).

We now add to this magnetically dated calendar of events the technique of isotope geochemistry, particularly the study of oxygen. Among its many other roles on Earth, oxygen is a constituent of the skeletal parts of carbonate-bearing plankton, the microscopic creatures that drift around the oceans. Some plankton build thin "shells" from calcium carbonate, and it is these that form the bulk of most deep-ocean sediment cores. Happily for us, and presumably for the plankton, there are two stable forms, or *isotopes*, of oxygen. The common isotope has an atomic weight of sixteen, whereas the less common one's weight is eighteen. Both isotopes react (in the same chemical proportions as they are found in nature) with other elements to form such chemical compounds as calcium carbonate. However, because of their different atomic weights, the *rates* of chemical reaction will differ. The same is true for strictly physical processes, such as the vaporization of seawater from the ocean surface. What this means is that in a reaction between seawater and a dissolved carbonate, the oxygen 18 isotope is preferentially left behind in the carbonate. Thus, the ratio of oxygen 18 to oxygen 16 will be higher in the carbonate plankton than in the parent seawater. Furthermore, studies of recent plankton from oceans of known temperatures show that the lower the ambient temperature, the higher the ratio of oxygen 18 to 16 will be. So this ratio is a reliable paleothermometer for measuring the temperature of ancient seas. This works if there has been no concurrent change in the ratio of the two isotopes in the ocean reservoir itself, and no secondary chemical alterations.

When oxygen in the form of water evaporates from the ocean surface, oxygen 18 is again preferentially left behind. Under normal conditions, this is no problem because the evaporated seawater is eventually returned to the ocean with no net change for the reservoir overall. But if there is a disequilibrium in this great cycle, and the lighter evaporation product (oxygen 16) is sequentially stolen away and locked in ice sheets, then the isotope ratio of the ocean reservoir will increase. As the ice sheets melt, it will decrease.

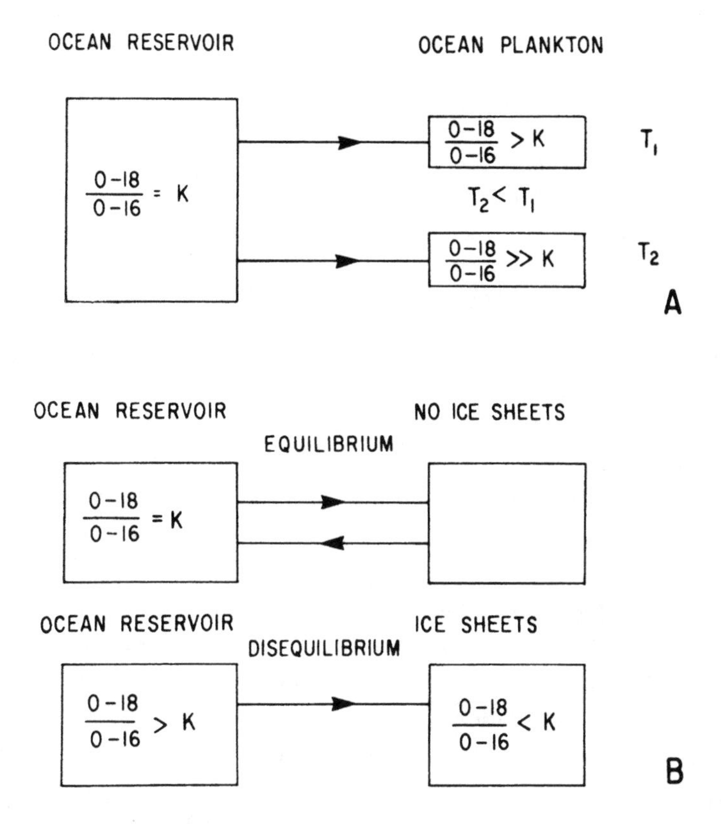

Oxygen-isotope relationships. (**A**) Dependence of oxygen-isotope ratio in carbonate-bearing plankton on ambient ocean temperature. (**B**) Dependence of oxygen-isotope ratio in ocean reservoir on continental ice sheet volume.

Thus the oxygen-isotope ratio is a reliable indicator of the waxing and waning of the glacial ice sheets over a period of time that is known through paleomagnetic dating. Each ice sheet was of comparable magnitude, building to a maximum thickness over an extended period of time and melting over a shorter period. A cycle of around 100,000 years for each glacial and interglacial stage emerges from these data and, given the isotope plot, it appears that we are within some 10,000 years of entering the next ice age.

Today we are wrestling with an even more complicated question—the causes of the Ice Ages. We have to look at three sorts of phenomena: (1) global climatic conditions persisting long enough to permit the ice sheets to form; (2) external events forcing variations in the amount of solar energy affecting the Earth; and (3) the internal dynamics of the ice sheet-atmosphere-ocean system.

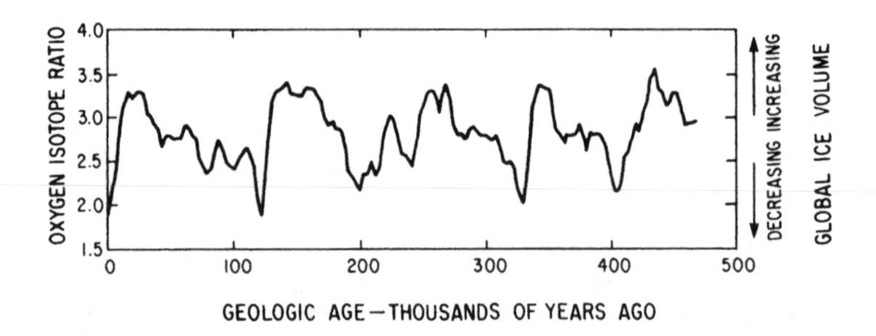

Variations in glacial ice sheet volume as inferred from oxygen-isotope measurements versus geologic time. From Imbrie and Imbrie, 1979.

The component of external forces has received the greatest attention from scientists. These forces are a matter of astronomy, concerning the distance between the Earth and the Sun and also the angle at which solar radiation strikes the Earth. The Earth revolves around the Sun in an elliptical orbit known as the plane of the ecliptic. The Earth also rotates around its own axis, which defines the equatorial plane. The plane of the ecliptic and the Earth's equatorial plane are set off from each other at an angle called the tilt. Both the shape of the elliptical orbit of the Earth around the Sun and the tilt between the ecliptic and equatorial planes vary slightly over time, given the perturbing gravitational effects of the other planets in the Solar System. The orbital variations have a period of 100,000 years; those of the tilt have a period of 40,000 years. In addition, the rotational axis of the Earth itself precesses, or wobbles, with a period of 22,000 years. The wobble comes about because the Earth is not a perfect sphere but instead an oblate ellipsoid, being wider (at the equator) than it is high (from pole to pole), and because of the Moon's gravitational attraction on this ellipsoid shape. These three astronomical variations are called the *Milankovitch cycles*, named after the Yugoslavian mathematician Milutin Milankovitch, who made the necessary, tedious calculations in the early 1900s before the computer era.

Each of these astronomical cycles causes seasonal variations in the amount of solar radiation reaching the Earth. It is possible to produce a model of this system, including the inferred volume of ice in a given ice age (based on the oxygen-isotope record) and to analyze it spectrally. And the model nicely shows the three periods of 100,000, 40,000, and 22,000 years. The only trouble is that the isotope record suggests that the longest period of the three is the dominant one, while the shortest period is the least significant. But from the astronomical calculations one would expect the reverse. The wobble period of 22,000 years should be the most potent, followed by the 40,000-year cycle of tilt, followed weakly by the 100,000-year period of orbital variation. This unfortunate discrepancy is well known to students of the Ice Ages and has forced them to look elsewhere for a resolution—to the internal dynamics of the ice sheet–atmosphere–ocean system, a complex dance in which the simple equations defining the mass of the ice balance of ice sheet and the global heat balance form a set of coupled, nonlinear equations: Chaos again.

Oceanographer Eric Posmentier studied the numerical response of this simple model—ice mass and heat—and found that there are three equilibrium positions for the system. If the ambient (surrounding) global temperature is sufficiently high, the world remains ice free. If it is sufficiently low, the world is completely covered by ice. During times of intermediate temperatures, the world goes through glacial–interglacial oscillations. Quite evidently, we are now in a glacial–interglacial situation, and whether or not the oscillations are driven by external forcing phenomena, the system responds with an alternation of glacial–interglacial regimes with a period of about 100,000 years.

The ambient global temperature is obviously an important condition in this dance. Ice sheets have been a prominent phenomena over the past one million years, but insignificant in the geologic record over the past 600 million years. Periods of glaciation account for a mere 1 percent of that unimaginably long time. We know from our isotopic studies of the ocean that the global temperature has steadily decreased by about 15 degrees Fahrenheit over the past 100 million years, apparently putting us in a temperature regime in which the "chaotic" cycles of glacial and interglacial epochs can thrive. It may be that, if the global temperatures continue to decrease over the next 100 million years, the Earth will pass into another equilibrium stage in which it will be encased in ice.

We know about the enormous decrease of 15 degrees in the planet's ambient temperature through plots of the oxygen-isotope ratio for plankton that lived near the ocean's surface at various equatorial sites. The ocean surface temperature is a usable proxy for the atmospheric temperature. It was already evident from fossil flora and fauna in Cretaceous times, 100 million years ago, that tropical climates existed over much of the world; the isotope studies simply provide a quantitative measure.

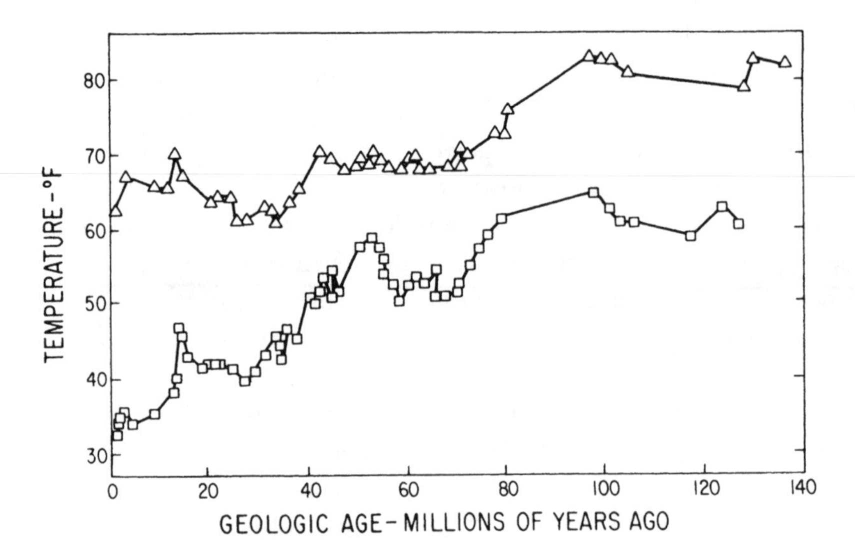

Surface and bottom temperatures inferred from oxygen-isotope measurements for low latitudes near the equator in the Pacific Ocean versus geologic age. Surface temperatures shown as Δ and bottom temperatures as ❑. From Berger and Crowell, 1982.

These same studies, performed on plankton living on the ocean's bottom, show that bottom temperatures changed even more, decreasing by some 30 degrees Fahrenheit over the same period. Today, the bottom waters of the world's oceans originate at the polar latitudes and, being colder (near the freezing point, 32 degrees Fahrenheit) and thus denser, they sink to the bottom and flow toward the equator. During the Cretaceous period, things were different. Presumably a competing source of bottom water existed, consisting of warm and more saline water (which was dense because of the load of salt), and this dominated the polar source. The ancient oceans would have been stratified as they are today, but the principal agent for stratification would have been salinity, not temperature, with the heavy, more saline strata at the bottom. Something had to have happened to start the oceans on the roller coaster of decreasing temperatures, and this could have been a shift in the nature of the atmosphere.

Two components of the atmosphere that concern us most, and for good reason, are oxygen and carbon dioxide. It is now possible to make reasonable estimates of how these gases have waxed and waned over geologic time in the envelope of atmosphere. In the long reaches of geologic time,

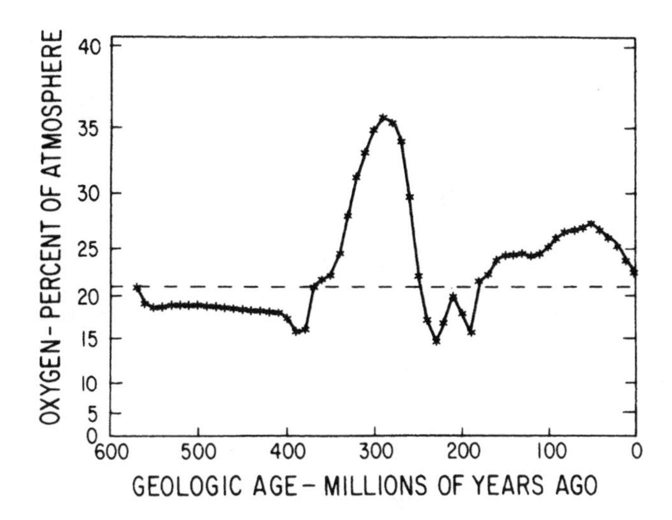

Plot of oxygen expressed as a percent of the atmosphere versus geologic age. From Berner and Canfield, 1989.

the reservoir of atmospheric oxygen has varied according to the amount of organic matter that is buried in sediments, as well as by the converse process by which organic matter in sedimentary rocks is weathered. At the same time, in photosynthesis (the process of building plant material), carbon dioxide is taken from the atmosphere and oxygen is released, generally with a balance between the two processes. But if there is a net burial of organic material over its decomposition, atmospheric carbon dioxide is depleted, and oxygen increases.

In the Carboniferous period, 300 million years ago, a major change in atmospheric oxygen took place. At the time, oxygen constituted 35 percent of the atmosphere, as opposed to today's 21 percent. This was the time when vast carbon-rich swamps came into being, visible today as great coal deposits such as those in eastern and central North America. The usual carbon-recycling process was disrupted as enormous quantities of organic material sank into the swamps (to eventually become our depletable fossil fuel sources). As a result, there was a huge net gain in oxygen over carbon dioxide released into the atmosphere.

Geologically, atmospheric levels of carbon dioxide follow a more complex cycle than oxygen. They depend on a number of factors: sedimentary burial of organic matter and carbonates; continental weathering of these

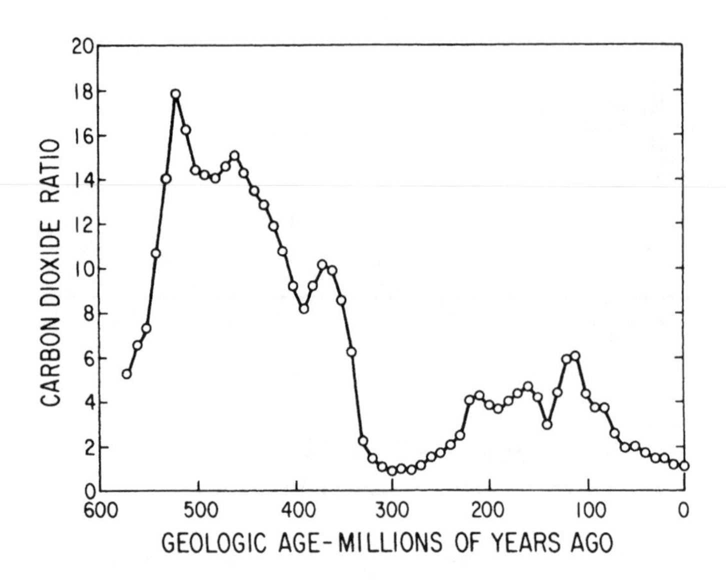

Plot of the ratio of carbon dioxide in the atmosphere in the past to that in the present atmosphere versus geologic age. From Berner, 1990.

same materials plus silicates; and the degassing of carbon dioxide from volcanoes and metamorphic events in the Earth's crust. Using a model that incorporated these processes, Robert Berner of Yale University recently estimated the levels of atmospheric carbon dioxide through time and found that 100 million years ago, they were *six* times higher than they are today. They were even higher earlier in geologic time. But in Carboniferous times, 300 million years ago, carbon dioxide levels in the atmosphere were at a low point, about the same as those in the Earth's atmosphere today.

We begin to see, then, what may have brought about the fifteen-degree drop in global temperature that began in Cretaceous times. The greenhouse effect of higher carbon dioxide levels has declined—as has the temperature—in an orderly fashion to today's level and today's temperature. As a corollary to this, besides what we might want to think of as "our" Ice Ages, the only other significant period of glaciation throughout geologic time occurred during the Carboniferous era, when carbon dioxide in the atmosphere was, like our era, at its lowest.

6

...And on Rare Occasions There Are Changes in Its Community of Living Things

When European settlers arrived on the east coast of North America, they were confronted by vast stretches of virgin forest made up largely of enormous old trees—the stuff of picture postcards. Today we rhapsodize over such places and seek to protect them, but to the early settlers fresh from the domesticated landscapes of Europe, this forest was a dark place, a savage wilderness that needed taming. And, practically speaking, none of it is left. Attitudes about "nature" change. Indeed, one scholar found that the very word "nature" has more than 100 separate meanings. To a geologist, nature is every thing and every process, and life is merely one phenomenon that has come about in nature. And life is dangerous. The existence of (1) the immune system and (2) the fossil record attests to this obvious but often overlooked fact of life. Immune systems, which all higher animals and most plants possess in one form or another, are in a nearly constant battle against other forms of life. The fossil record makes it clear that far more species have vanished than now live on the planet.

There are any number of "natural causes" for the death of an individual, but this phrase usually means that one or another "disease" has overcome the individual's immune system. It happens all the time. Throughout the long history of life on this planet, thousands of species have died out—also of natural causes. We don't think of such events as global or even regional catastrophes but merely the quiet winking out of local life forms. But when many individuals die suddenly, or when a huge number of species rapidly become extinct...the event is a catastrophe and commands our attention.

Beyond humanity's beastliness to its own kind, the greatest mass killers of people have not been the geological vagaries of the Earth, earthquakes and volcanoes and the like, but widespread epidemics, called *pandemics*. We tend to think of pandemics as things of the past, given our advances in medical science, but of course they are all too present. Not so long ago, in the influenza epidemic of 1918, more Americans died in a few months than did American military personnel in all the preceding years of World War I. Worldwide the total deaths from this influenza epidemic are estimated to be twenty million. As of 2007, an estimated 33 million people are infected with the AIDS virus, and the Joint United Nations Programme on HIV/AIDS and World Health Organization estimate that every day more than 6,800 persons become infected with HIV. The population of African nations is being devastated by AIDS, with sub-Saharan Africa accounting for almost one third of all new HIV infections and AIDS deaths globally in 2007.

The greatest pandemic scourge in recorded history was the plague, of which there were two major eruptions: the Justinian plague of the sixth century, and the bubonic plague of the fourteenth century. In each of these pandemics, one-half the population of the Western world perished, not to mention untold millions of people in Asia and the Middle East. No other environmental event can compare with these two outbursts of disease in the severity of their effects on humanity.

Under normal conditions, plague bacilli are present in low levels in many wild rodents. They are still with us, too, in a relatively dormant state, centered in eastern Asia and other smaller locales, including the American Southwest. There are occasional cases of plague in these regions when fleas, which typically transmit the bacilli from one rodent to another, pass them on to a human. No one knows just what circumstances lead to a sudden outburst of epidemic potential, but in such times the plague bacillus population increases enormously. When a flea bites an infected rodent, the flea ingests the bacilli, which replicate in its digestive system to such a degree that they block the insect's gut. Unable to assimilate its host's blood, the flea becomes ravenously hungry and repeatedly bites the rodent, each time injecting more bacilli into the rodent's bloodstream. If the host rodent dies, which it usually does, the flea moves on to the next available rodent. But as the number of rodents decreases, the flea seeks other hosts, including humans. Thus an epidemic starts.

There are three varieties of plague: bubonic, pneumonic, and septicemic. By far the most common, bubonic plague is transmitted by the rat flea. After a few days of incubation in a human, the lymph nodes in the armpit, groin, or neck enlarge, followed by internal hemorrhaging. This produces purple to black blotches called buboes on the skin, giving the disease its name. Their color provides the terrifying name of the fourteenth-century pandemic, the Black Death. Severe disorders of the nervous system soon follow, ultimately leading to death in 50 to 60 percent of the victims.

Pneumonic plague is a derivative form, unusual in that it can be transmitted directly from human to human without the mediation of a flea. Instead, the infection lodges in the lungs, leading to severe coughing and airborne transmission. It is more lethal than bubonic plague, causing death in almost every case. Septicemic plague, which is rare, is also transmitted by the rat flea and leads to massive numbers of the plague bacillus in the bloodstream. Death occurs in all cases within a day of infection.

The Black Death had its beginnings with an eruption of the disease in the Gobi Desert in the 1320s, where it infected the nomadic Mongols whose horsemen and supply caravans spread it throughout the rest of Asia. China and the entire Islamic world, along with the Christian portions of the eastern Mediterranean, were devastated: between one-third and one-half of the populations in these regions perished. The Black Death reached Europe in October 1347, on board a Genoese trading fleet that anchored at Messina, the main port of Sicily. The crews were sick and dying and the fleet was quickly quarantined—no deterrent for the rats on board, however. By November all of Sicily was stricken; by December it had spread through southern Italy. The rats traveled as supercargo on coastal trading ships and with overland traders. By January 1348, it reached southern France and by spring, Paris. It crossed the English Channel in September, and, from the British Isles, it spread to Scandinavia. Urban and coastal communities with their large rat (and human) populations were most severely hit, the more remote rural and mountain towns less so.

The first cycle of the Black Death lasted five years, from 1347 to 1352, followed by succeeding episodes through the remainder of the fourteenth century and into the fifteenth. The death toll in the first five years has been estimated at twenty-three million, with succeeding cycles carrying off another twenty million lives. The toll of forty-three million amounts to about half the population of the Western world at that time, and such devastation and the resulting demoralization defy the imagination. A sense of this tragedy can be found in some contemporary accounts, excerpted here

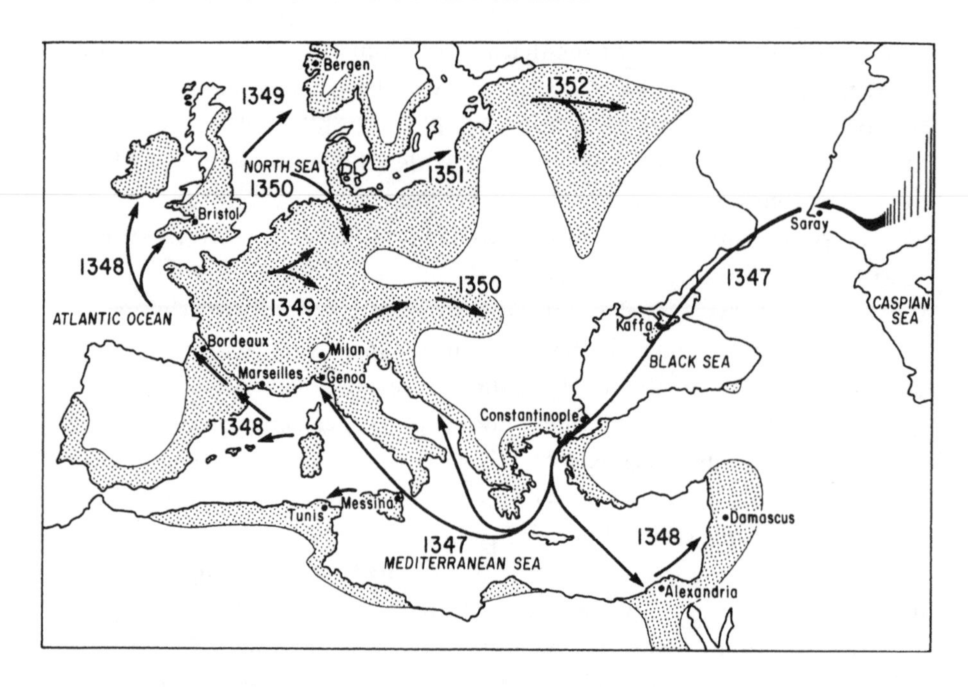

The Black Death came from central Asia to Europe via the silk road, arriving in Kaffa about 1347. From there it was carried by ship to the major ports of Europe and northern Africa. Most of Europe was affected before the epidemic finally subsided in 1352. Milan, the largest city to escape the plague, is believed to have done so because it is the Italian city farthest from the sea. Adapted from McEvedy, 1988.

from Robert Gottfried's book, *The Black Death*. According to chronicler Agnolo di Tura:

> The mortality in Siena began in May. It was a cruel and horrible thing; and I do not know where to begin to tell of the cruelty and pitiless ways. It seemed that almost everyone became stupefied seeing the pain. It is impossible for the human tongue to recount the awful truth. Indeed, one who did not see such horribleness can be called blessed. The victims died almost immediately. They would swell beneath the armpits and in the groin, and fall over while talking. Father abandoned child, wife husband, one brother another; for this illness seemed to strike through breath and sight. And so they died. And none could be found to bury the dead for

> money or friendship. Members of a household brought their dead to a ditch as best they could, without priest, without divine offices. Nor did the death bell sound. In many places in Siena great pits were dug and piled deep with the multitude of the dead. And they died by the hundreds, both day and night, and all were thrown in those ditches and covered with earth. And as soon as those ditches were filled, more were dug. And I, Agnolo di Tura...buried my five children with my own hands....And so many died that all believed it was the end of the world.

Another close observer was Giovanni Boccaccio, who is best known for *The Decameron*, an early collection of bawdy and irreverent tales "related by an honourable company of seven ladies and three young men made in the days of the late deadly pestilence." Indeed, the ten raconteurs had escaped from the city of Florence to the relative safety of the countryside. Like jokes in wartime, the tales were a way to keep one's mind off the horror. Boccaccio wrote that in 1348,

> the deadly plague broke out in the great city of Florence....Whether through the operation of the heavenly bodies or because of our own iniquities, which the wrath of God sought to correct, the plague had arisen in the east some years before, causing the death of countless human beings. It spread from one place to another until, unfortunately, it swept over the west. Neither knowledge nor human foresight availed against it, though the city was cleansed of much filth by chosen officers in charge and sick persons were forbidden to enter it, which advice was broadcast for the preservation of health. Nor did humble supplication serve. Nor once but many times there were ordained in the form of processionals and other ways of propitiation of God by the faithful, but in spite of everything, towards the spring of the year the plague began to show its ravages in a way just short of miraculous. It did not manifest itself as in the east, where, if a man bled at the nose he had certain warning of inevitable death. At the onset of the disease, both men and women were affected by a sort of swelling in

> the groin or under the armpits, which sometimes attained the size of a common apple or egg. Some of these swellings were larger and some were smaller, and all were commonly called boils. From these two starting points the boils began in a little while to spread and appear generally all over the body. Afterwards, the manifestations of the disease changed into black or lurid spots on the arms, the thighs and the whole person. In many ways, these blotches had the same meaning for everyone on whom they appeared.... Such was the cruelty of heaven and to a great degree of man that between March and the following July it is estimated that more than 100,000 human beings lost their lives within the walls of Florence, what with the ravages of the plague and the barbarity of the survivors towards the sick.

In England, Henry Knighton of Leicester noted the progress of the disease: "The dreadful pestilence made its way along the coast by Southampton and reached Bristol, where almost the whole strength of the town perished, as it was surprised by sudden death; for few kept their beds more than two or three days, or even half a day. Then this cruel death spread on all sides, following the course of the sun. And there died at Leicester, in the small parish of Holy Cross, 400; in the parish of St. Margaret's, Leicester, 700; and so in every parish, in a great multitude."

The bishop of Bath and Wells, concerned about the penitential state of his parishioners, wrote as follows to his parish priests:

> The contagious pestilence of the present day, which is spreading far and wide, has left many parish churches without parson or priest to care for the parishioners. Since no priests can be found who are willing, whether out of zeal or devotion to exchange for a stipend, to take pastorale care of these aforesaid places, nor to visit the sick and administer to them the sacraments of the church, we understand that many people are dying without the sacrament of penance...persuade all men, in particular, those who are now sick or should feel sick in the future, that, if they are on the point of death and cannot secure the services of a priest, then they should make confession to each other...or if no man is present, then even to a woman.

Because it was commonly believed that the plague was a sign of God's wrath at sinful man, long chains of penitents roamed the countryside, whipping and abasing themselves, and praying to God for mercy. The dragon that is borne aloft is a symbol of the Great Tempter himself. Metropolitan Museum of Art, New York.

Equal rights were not high on the bishop's list of priorities.

As with the Lisbon earthquake, the public was convinced that the plague was a curse from God. An unusual form of atonement arose in response: the flagellants, who roamed from community to community in central Europe. When a group of flagellants entered a town, they made their way to the most prominent church, stripped to the waist, and formed a circle. They would extend their arms in the form of a cross and receive a lashing from their fellows in celebration of the Passion of Christ. People approvingly turned out in great numbers to watch the flagellants and listen to their anticlerical preachings. Before long they had locally usurped some Church functions, such as granting—and charging for—absolutions. But they had gone too far, and in October 1349, Pope Clement VI issued an edict condemning flagellism. The practice was soon repressed but an evil effect lingered on. The flagellants were not merely anticlerical, but antisemitic. Jews became scapegoats for the plague: in the spring of 1349 the Jewish community in Frankfurt was destroyed, followed by those in Mainz and Cologne. By 1351, more than 200 Jewish communities had ceased to exist, and more than 300 separate massacres had taken place. In the latter part of the century, Jews migrated en masse from central to eastern Europe. The holocaust of World War II had a precedent in the plague-ridden fourteenth century.

The ramifications of the plague were, of course, innumerable. Historians probably are justified in claiming that it marked the end of medieval Europe and the beginning of the modern age, an upheaval that saw a historic shift of power from the nobility and the Church to the merchant and agrarian classes.

No one knows what conditions caused the plague bacilli to explode from their normal dormant state in the Gobi Desert rat and flea population, but there is a better understanding of why the plague finally died out. With so many human beings dead, there were fewer pathways for the propagation of succeeding plague cycles. And that portion of the human population that had contracted plague but had survived probably developed an immunity to further attacks. This is borne out by the fact that the succeeding cycles that went on throughout the fourteenth century and beyond wreaked the most havoc on the young, people born after the initial cycle of 1347 to 1352. There is also the possibility that the plague bacillus, *Yersinia pestis*, developed over the years into a subsidiary, nonvirulent form, *Yersinia pseudotuberculosis*. That is, *most* of the bacilli did. As noted, the plague bacillus still exists.

In such ways does nature rise up and suddenly take its toll of vast numbers of people: no inorganic object or event has been more destructive of human life. But from the viewpoint of the nonhuman creatures of the world, humankind has been the plague. We have been responsible for virtually all the recorded extinctions of species in modern times and may well have put still other species over the brink in our infancy as a race. In any event, there are few modern cases of extinction of animal species that are clearly related to climatic changes or other geologic or "natural" causes. Of the 4,200 species of mammals, an estimated 63 have become extinct since A.D. 1600, all at the hands of humans. The same is true for birds: of 8,500 species, 88 have been extinguished in the same interval. We have accomplished this in several ways.

Prominent among them is outright overkill, for food, pleasure, trophies, or riches. Overvigorous hunting is the sole reason that marine mammals like Stellar's sea cow, the Atlantic gray whale, the Japanese sea lion, and the Caribbean monk seal have been exterminated. Overkill is the chief cause for the modern extinctions of forty-six large terrestrial mammals, as well as for 15 percent of bird extinctions, including the great auk, the passenger pigeon, and the dodo.

Predators accidentally or deliberately introduced into new places—especially that ubiquitous human companion, the rat—have had devastating effects, particularly on island bird populations. Hawaiians traveled to Polynesia bent on obtaining feathers for ceremonial purposes. The rats they inadvertently brought with them from Polynesia carried diseases that wiped out several species of Hawaiian birds. More subtle are introduced *competitors*, like the European starling, which was brought to New York City in the late nineteenth century and exploded outward, usurping the habitat of such birds as purple finches and eastern bluebirds, reaching by the mid-1900s into virtually every county in the United States, with an estimated population of 600 million.

Introduced diseases, from Dutch elm disease to smallpox, have devastated if not eliminated entire populations of plants and animals, including humans. A number of Indian tribes in the New World were decimated by the ravages of smallpox, a disease of the Europeans to which native peoples had no immunity. (At the same time, there is some evidence that syphilis did not occur in Europe until Columbus's expedition returned to the continent.)

The human destruction of habitat has been an important factor in the extinction of many continental land birds, and this rapidly accelerating process bids fair to bring about one of the greatest and most abrupt extinction episodes in all of geologic time. How much habitat has been turned from its wild origins to human use? According to the World Resources Institute, approximately 40 percent of all the photosynthesis occurring on Earth is expended to produce energy or goods for humanity. It is estimated that Minnesota-sized tracts of rain forest are being lost each year to human activity, and there are more species of plants and animals in rain forests than in any other habitat, often local species, which are quickly and easily snuffed out. Often overlooked as well are other habitats such as savannas, which are disappearing at alarming rates, along with species dependent on them. Species are leaving this plane of existence at an estimated rate of one a day, and the fact that most are nameless tropical insects and plants is hardly cause for sanguinity.

There is a school of thought that we have been extinguishing species for a long time. It goes under the rubic of Pleistocene overkill, the theory being that, in North America at least, the bands of people who crossed the Bering Strait and populated this hemisphere killed off the last populations of the big animals, not only by the use of their arrowheads and spear points but by running vast herds of creatures over cliffs. Images of a mastodon towering over a frantically determined band of paleohunters armed with little but crude spears have a place in popular imagination. Nothing (save dinosaurs) in the fossil record captures the imagination more than these great beasts and the fauna associated with them—cave bears, giant sloths, and saber-toothed cats, the eerily familiar beasts, many of which so helpfully interred themselves in the La Brea tarpits in Los Angeles for later human investigators. Their demise (in some instances, at least) does correspond with the rise of prehistoric human hunters as a distinct ecologic entity at the end of the last glacial period. But this, too, was a time of great and rapid change in climate and vegetation. It is unlikely that humans can take credit for annihilating such animals as saber-toothed tigers, but they might be implicated in an accessory role in the end of the mastodon and the wooly mammoth.

The American mastodon of Central and North America and the woolly mammoth of Eurasia and western North America (with especially large populations near the ice sheet edge in Siberia) were the last of the large Ice Age mammals to roam the Earth. Until the eighteenth century, no one had even imagined the existence of these large creatures, nor had they postulated the Ice Ages. One of the early descriptions of mastodon remains appears in a letter to the Reverend Cotton Mather from Massachusetts Governor Joseph Dudley, who was presented in 1706 with an immense tooth and several enormous bones found along the banks of the Hudson River. But for its size—six inches long, thirteen inches in circumference, and two pounds in weight—the tooth resembled a human eyetooth, and Governor Dudley concluded that it was indeed human having belonged to one of the races of giants mentioned in the Bible. To the Reverend Mather he wrote:

> The distance from the sea takes away all pretension to its being a whale or animal of the sea, as well as the figure of

> the tooth. Nor can it be the remains of an elephant; the shape of the tooth, and ad-measurement of the body in the ground will not allow that.
>
> There is nothing left but to repair to those antique doctors for his origin, and to allow Dr. Burnet and Dr. Whiston to bury him at the Deluge; and if he were what he shows, he will be seen again at or after the conflagration, further to be examined.

At about the same time, discoveries of mammoth remains in Eurasia were brought to the attention of Peter the Great of Russia, who was something of a Renaissance man. The Czar was intrigued and accepted the gift of a mammoth tusk from Basilius Tatischow, director of mines in Siberia. More than merely curious about such things, Peter issued a directive that was later referred to in a scientific article of 1728 by Hans Sloane, then president of the Royal Society in London: "It is to be hoped that this matter will one time or another be set into a still clearer light, particularly after the order of his late Czarish Majesty was pleased to give to the Governor General of Siberia, to spare no care nor cost to find a whole skeleton of this animal, and to send it to Tatischow."

Depiction of the woolly mammoth. Field Museum of Natural History, Chicago.

Depiction of the American mastodon. American Museum of Natural History, New York.

With such prodding, more mammoth specimens—and in North America, more mastodon remnants—began to come to light. In 1766, while he was resident in London, Benjamin Franklin received a collection of mastodon bones that his colleague George Croghan had found in Ohio. The teeth belonged to a meat-eating animal, Franklin thought (as did many of his contemporaries), but he later changed his mind, agreeing with yet other students that the mastodons were too heavy and therefore slow to prey on other animals. The teeth, it was felt, would serve just as well for chewing branches and shrubs as they would for a life of flesh eating. What puzzled Franklin was that the bones were found at latitudes too cold for modern elephants. In a letter to Croghan he opined that the climate of North America might have been altered by a shift in the Earth's axis of rotation—an inaccurate supposition for the actual timing of the Ice Ages, as it happens, but an imaginative suggestion, one that borders on the concept of continental drift, wandering poles being related to the motion of the plates and continents.

Another American of many interests, Thomas Jefferson was also fascinated by these discoveries and encouraged the search for such remains in his native Virginia. That search succeeded when workmen excavating for saltpeter in a cave came upon some curious animal bones. They were quite different from the mastodon bones from the Hudson and those that Croghan

Thomas Jefferson (1743–1826). American Philsophical Society, Philadelphia.

had found, and demonstrated that another huge animal had once existed in North America. In 1799, Jefferson published a paper in the *Transactions of the American Philosophical Society*, naming the new animal *Megalonix* (great claw) and conjecturing that it was a huge lion. At the time, Jefferson was serving as John Adams's vice-president, and this is the only known incident of a U.S. vice-president publishing a scientific article while in office.

Jefferson was a strict adherent to the principle of continuance and uniformity—that is, things are now as they have been in the past. Just as he dismissed the notion of meteorites from outer space striking the Earth, he did not believe that either evolution or the extinction of species could occur. In a previous publication, he had written: "Such is the economy of nature, that no instance can be produced, of her having permitted any one race of her animals to become extinct; of her having formed any link in her great work, so weak as to be broken." He continued his interest in such matters while president from 1801 to 1809, instructing Meriwether Lewis and William Clark, on the famous expedition that opened up the American West, to search for such remains, an important contribution to the emergence of the science of vertebrate paleontology.

The *Megalonix jeffersoni* reconstructed by modern paleontologists. American Museum of Natural History, New York.

The demise of the mammoths and mastodons and their cohorts at the end of the last Ice Age was not an extensive event compared to many mass extinctions in the geologic record. It involved chiefly very large animals that existed in small populations—that is, there were few individuals in any of the species affected—and small populations are always at greatest risk of extinction. Their demise took place over an extensive period from about 20,000 to 8,000 years ago, meaning that it was not the result of any catastrophic event, either terrestrial or extraterrestrial in origin. Disease might have played a role, as could inbreeding (the silent stalker of all relatively small populations, leading to reduced fertility and increased infant mortality), but such effects cannot be judged readily from skeletal remains, nor has evidence been found in the very few frozen carcasses that have been recovered from the ice.

Two causes for these animals' extinction are most often cited: climate change and human predation. They disappeared during the beginning of the present interglacial period when, as the ice sheets retreated, the climate changed rapidly, as did the nature of the vegetation in this vast subglacial region. Ben Franklin's second theory regarding the eating habits of mastodons was, of course, correct: the mastodons and mammoths were herbivores. In Siberia, with the interglacial rise of the sea level and the flooding of great tracts of land, the grasslands of the Arctic continental shelf disappeared, and much mammoth habitat was removed. Nearby, the cold tundra of Siberia became boggy and less habitable for these great beasts. In North America, the spruce trees of the Ice Age south, which served as mastodon food, died out, to be replaced by species we now think of as southerly. The remaining mastodon populations appear to have been centered in the northeast. There, spruce woodland and tundra were replaced by pine (also a good food source) and then in turn replaced by deciduous species. As the pine forests were reduced to islandlike patches, the mastodons were effectively restricted to them, living in small groups that typically tended to decline on their own. When these islands themselves disappeared, so did any remaining mastodons.

It is conceivable that humans may have provided a late *coup de grâce* to mammoth and mastodon populations in the more temperate latitudes. Cave paintings in Europe show mammoth hunts, and there is definitive evidence of a mastodon kill at the edge of a former water hole in Venezuela 13,000 years ago. But changing climate was chiefly responsible for the ultimate extinction of these Pleistocene beasts, attested to by the fact that in previous ice ages—before the evolution of humans—other associations of large animals became extinct with the onset of interglacial intervals.

Life has existed for 600 million years, far before any glacial period. During that immense period of time, there have been several major turnovers in the community of living things. From the beginning of geology as a science, these great paleontological turnovers have provided a convenient way of breaking up the long periods of time geologists must contend with. Thus we have the Paleozoic (literally, *ancient fauna*) and the Mesozoic (*middle fauna*) eras. These breaks between geologic periods are often shown graphically as horizontal lines across a vertical bar graph representing time: here begins the Triassic Age, and here it ends. Such representations on paper (where the thickness of a printed line itself has to represent 100,000 years or more) tend to make transitions from one age to another seem swift, clean, abrupt, and, presumably, catastrophic. And no transition seems to be more abrupt in most minds than the end of the dinosaurs.

Many of us have a picture in mind of dinosaurs as monstrously large creatures that once roamed a swampy Earth but then were suddenly gone. The question of why they vanished so suddenly has fascinated humans of all ages and persuasions since a fossil dinosaur bone was first recognized for what it was. This was the last major extinction crisis on Earth, occurring about sixty-six million years ago, bringing down with the dinosaurs a host of other kinds of creatures, mostly marine plankton and shellfish. And it all happened in what amounts to a geologic instant. In fact, the extinctions took place over a period of time we can barely imagine—thousands to hundreds of thousands of years. Writing originated 7,000 years ago and our human experience of changes on Earth can be measured in tens, perhaps hundreds, of years. From our own limited experience of environmental change, it is nearly impossible to appreciate, even to grasp, the kinds of changes that have taken place over eons. Imagine, then, the excitement of finding a distinct line in nature that corresponds to the distinct line in the geologic timetable: the moment the dinosaurs vanished, the transition between Cretaceous times and Tertiary times, known specifically as the Cretaceous/Tertiary transition (in shorthand, the K/T boundary). Imagine the excitement of finding a single event that could have brought down that vast race of terrible lizards, a single cause for such a great dying.

Scientists have long speculated about the extinction of the dinosaurs, sometimes coming up with explanations as marvelous as the animals themselves. At one point, it was suggested that the dinosaurs might have died out from constipation because, as flowering plants (angiosperms) became dominant over evergreens (gymnosperms), the oil in angiosperm seeds would have blocked dinosaurian digestive systems. Such interest on the part of paleontologists and others had tottered along, a minor theme in science, the majority satisfied generally with the notion that some complicated changes in climate and environment would have led to the dinosaurs' end. Then, in 1980, a physicist named Luis Alvarez came forward with the hypothesis that this last great extinction was caused by the impact of a gigantic asteroid. The impact would have created a huge dust cloud that shrouded the Earth, curtailing if not cutting off photosynthesis and leading to the demise of those animals dependent on plants—such as herbivorous dinosaurs, followed promptly by the carnivores.

Alvarez was not given to wild speculations. He was a Nobel Prize winner in physics and a highly respected scientist. When he spoke, people listened. He could point to a smoking gun in the geologic record. His son, geologist Walter Alvarez, and others had discovered in rock strata located in Italy, Denmark, and New Zealand and dated precisely to the Cretaceous/Tertiary boundary a thin layer bearing anomalous amounts of a rare element, iridium. Iridium, a metallic element akin to platinum, is extremely rare in the Earth's crust. But in the widespread, presumably worldwide layer Alvarez pointed to, it occurred in the range of five to ten parts per billion (ppb)—an iridium bonanza that could only be explained as having come from an extraterrestrial source, since iridium is far more common in meteorites. (For a discussion of meteorites, see Chapter 4.)

The Alvarez hypothesis was simple, straightforward, and dramatic, and it was argued by an eminent scientist. As Australian journalist Ian Warden put it in the May 20, 1984, issue of *The Canberra Times*: "To connect the dinosaurs, creatures of interest to but the veriest dullard, with a spectacular extraterrestrial event like the deluge of meteors... seems a little like one of those plots that a clever publisher might concoct to guarantee enormous sales. All the Alvarez-Raup theories lack is some sex and the involvement of the Royal family and the whole world would be paying attention to them."

Indeed, practically the whole world did. Dinosaurs made the cover of *Time*. And in the interval, it is fair to say, virtually all interested lay persons are satisfied that the answer to the dinosaurs' extinction is now known: a catastrophic global nuclear-type winter brought on by extraterrestrial impact, or impacts. But geologists had over the past century painfully rid themselves of the previous school of catastrophism, and to them this idea smacked of neocatastrophism. Furthermore, the impact theory left many geologic and paleontogical matters unaddressed. And so the battle was joined.

The vertebrate paleontologists said the geologic record showed that the demise of the various dinosaur species occurred sequentially over 100,000 to 500,000 years, not in a relatively brief cataclysmic event that left the planet suddenly littered with dinosaur carcasses. The final disappearance in the fossil record, they said, of the nearly ubiquitous *Triceratops*, the last of the dinosaurs, occurred stratigraphically *below* (meaning earlier) the level of iridium-bearing clays. A few such cautionary facts would have been sufficient to scotch the asteroid-impact theory, but it had its own momentum by then. Many scientists had "invested" in it, the public loved it, even some elements of the "scientific press" loved it. (For example, *Science* magazine, the journal of the American Association for the Advancement of Science, appeared reluctant to publish any findings that gainsaid the impact theory.)

As has happened from time to time in the brief history of geology, matters became rancorous. Paraphrasing another, earlier physicist, Lord Kelvin, Luis Alvarez lashed out against paleontologists. In *The New York Times* of January 19, 1988, Alvarez said, "They're really not very good scientists. They're more like stamp collectors."

But a continuing collection of what we might call "geophilatelic insights" nonetheless raised questions about the impact theory. The theory calls for the creation of a dust cloud 1,000 times greater than that produced by the eruption of Krakatoa. But Dennis Kent of Columbia University pointed out that 75,000 years ago the eruption of the still larger volcano Toba sent up a dust cloud estimated to be at least 400 times that of Krakatoa. Yet no abnormal rate of extinction occurred after the Toba eruption.

While dinosaurs have naturally received the greatest attention during the controversy, particularly among the public (which may not even be aware that a controversy exists, so foregone a conclusion has the impact theory become in the press), the dinosaurs are not the most satisfactory group of

Depiction of *Triceratops*, last of the herbivorous dinosaurs, and *Tyrannosaurus*, last of the carniverous dinosaurs. Field Museum of Natural History, Chicago.

animals to examine when trying to determine the specifics of the extinction process. Only a relative few species of them existed at any given time during their 160 million years on the planet, and relatively few specimens have been preserved in the geologic record as fossils—and these are only found in isolated locales. The theory must explain why so many forms of life persevered: other reptiles, including big ones like crocodiles; the mammals; and those dinosaurian derivatives, the birds, to name but a few. It must explain as well the other extinctions of the age—the plankton and the shellfish. Here we are dealing with a large number of species, and a large number of specimens.

Paleontologist Gerta Keller looked at the extinction record of ocean plankton at the Brazos River in Texas and at El Kef in Tunisia, and found that the plankton species disappeared in a sequential, stepwise manner over a period of a few hundred thousand years. Other researchers have found the same extended, sequential extinction record for shellfish species in Antarctica

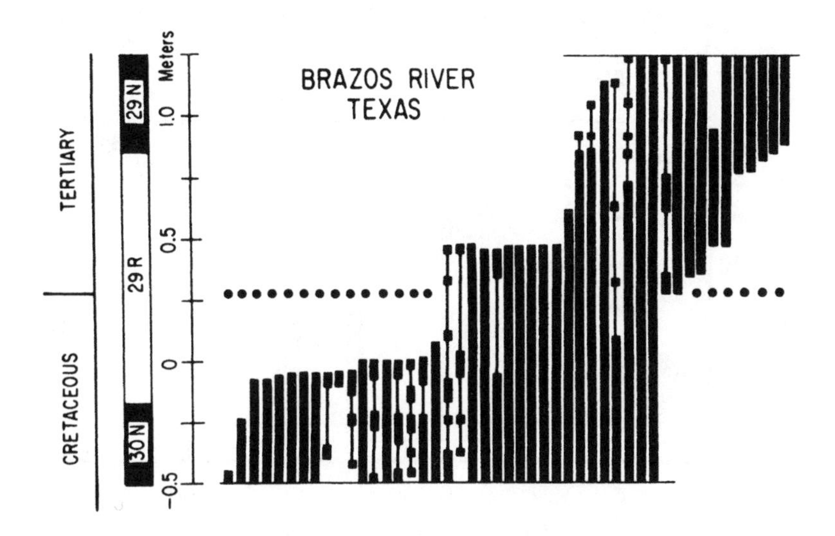

Foraminiferal extinction record across the Cretaceous/Tertiary transition at Brazos River, Texas. Magnetic polarity interval, 29R, shown to the left in the diagram, is approximately 500,000 years in duration. There are several iridium anomalies at the Brazos River section and the level of one of these is shown by the dotted line. Nearly half (46 percent) of the species disappear during the first extinction event; 33 percent disappear during the second extinction event. The remaining 21 percent of the Cretaceous species survive longer. The incoming, replacement Tertiary species are shown to the right in the figure. Adapted from Keller, 1989.

during this general time period. The onset of the shellfish extinction series also preceded that of the ocean plankton. Clearly, extinctions are not simple, instantaneous events. They are complex and occur over periods of geologic, not human, time (until recently). Such paleontological findings should be enough, in and of themselves, to rule out an asteroid impact as their cause. In a system as complex as the Earth and its living forms, it may be useful to

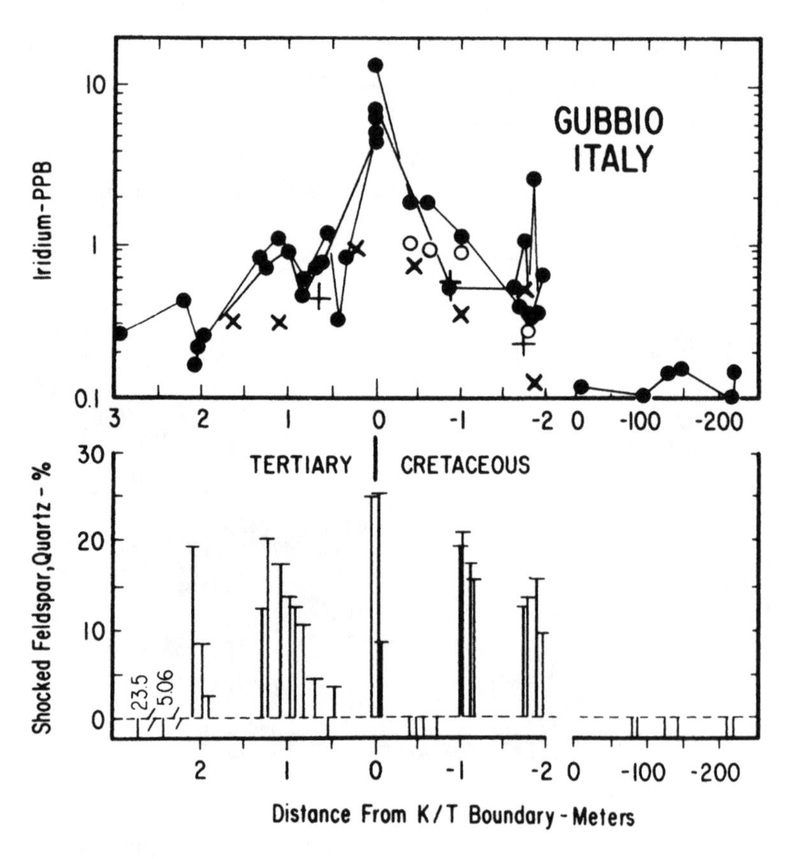

Comparison of iridium and shocked mineral results from two Cretaceous/Tertiary sections in Italy. The values to the right on the iridium graph are background iridium levels and the short vertical bars below the zero line on the shocked mineral graph indicate that samples at these locations show no shock features. Note that the iridium enhancements and shocked minerals extend over an interval of about four meters, straddling the iridium peak for a total of approximatley 400,000 years. Adapted from Crocket et al. (1988) and Carter et al. (1989). The additional data shown by open circles and crosses are checks on the iridium levels by other investigators.

keep in mind a remark H. L. Mencken made in another context: "For every complex problem, there is a solution that is simple, neat and wrong."

The proponents of the impact hypothesis were not dissuaded, and the controversy has gone on, an ancient detective story with many dead bodies and few clues to the cause of death. And those clues can be variously interpreted. If paleontological evidence is a hindrance to the impact thesis, its proponents do, after all, have the smoking gun: the iridium layer. Yet it is not conclusive proof of an extraterrestrial origin. Subsequent examinations of the iridium layer have been made by James Crocket of McMaster University in Canada and Robert Rocchia of the Centre National pour Recherche Scientifique in Paris and others, at the original sections in Denmark, Italy, and New Zealand, as well as elsewhere around the world. They have shown that the iridium does not appear as a single spike in the geologic record, which would be required for a single impact; instead, it is distributed over a few hundred thousand years, building to a peak and then subsiding. What might have caused such a phenomenon?

In the Deccan plateau region of western India is a vast landform known as the Deccan Traps, the result of the largest flood of molten basalt during the past 200 million years, with a volume of 300,000 cubic miles. Radiometric and paleomagnetic age-dating show that the Deccan event occurred at the time of the K/T boundary, over the relatively short (geologically speaking) period of a few hundred thousand years. At the same time, intense volcanism was taking place in western North America and the southeast Atlantic. It has recently been shown that iridium is relatively more common below the Earth's surface and is greatly enhanced in airborne particles from recent volcanic eruptions, such as Kilauea in Hawaii, Piton de la Fournaise on Réunion Island in the Indian Ocean, volcanic eruptions on the Kamchatka Peninsula in eastern Asia, and volcanic dust bands from Antarctic ice samples. Thus, the occurrence of iridium in geologic strata cannot be taken as categorical proof of an asteroid impact. Indeed, the observed results of Crocket and Rocchia are in accord with the record of Deccan volcanism, which built to a peak over 100,000 to 200,000 years and then, over the next 200,000 years, subsided.

Another problem the impact hypothesis faces is the lack of clear evidence of an impact crater. Several locations have been suggested, including Iceland, the Andaman basin in the Indian Ocean, Iowa, Hudson Bay, the Gulf of Saint Lawrence, the Colombian basin in the Caribbean, and the Chicxulub structure on the Yucatan Peninsula in Mexico. All, with the exception of the Chicxulub structure which is still under debate, have been subsequently ruled out as possible impact sites.

The advocates for terrestrial causes for the extinctions may seem to be party poopers, but the facts seem to lie on their side. And the facts may be more relevant, in many ways, to our present environmental concerns. For with the kind of volcanism that was unquestionably taking place over a few hundred thousand years at the K/T boundary, huge sulfur dioxide emissions would have led to widespread acid rain, carbon dioxide would have produced an increase in the greenhouse effect, and chlorine emissions would have depleted the ozone layer. These are our problems today, but they existed on a far greater scale in those days. Added to them was a major regression in the sea level, with its concomitant complications for global climate and habitat. With all of this going on, the need for an extraterrestrial explanation of the extinction events of the age becomes less urgent.

Whether Luis Alvarez's idea proves to be right or wrong—and the recent geologic and paleontological findings suggest that it has serious problems—he deserves a great deal of credit for reviving interest in one of the fundamental geologic problems: the causes of mass extinctions that have occurred over the past 600 million years. At the same time, an interesting lesson to do with the funding of science can be learned. Much of the basic research done in the United States depends on federal research dollars. Understandably, there is a lot of infighting when it comes to setting the priorities for how this federal money is to be divided. Planetary geology is funded for the most part by NASA. Most scientists agree that while NASA's funding for hardware and vehicles has been substantial, its funding for basic research has been inadequate over the past several years. The impact hypothesis provided an opportunity for NASA-dependent scientists to argue convincingly for increased funding. Thus, a great deal of scientific activity has been stimulated on both sides of the question, with a great deal of new information accumulating on the nature of the K/T extinctions.

We know far less about earlier mass extinctions. For one thing, they simply have not been studied as intensely. Also, the further back the geologic record extends, the less distinct and more obscure it is. Data are sparse and trying to relate dimly perceived events to a precise time is difficult. There have been recurrences of extinctions every twenty-five to thirty-five million years, but apparently no strict periodic repetitions. Some of the events have been regional, some global in extent. Most that we know of have involved marine animals and plants, but that is because the fossils of marine plants and animals are often all that remain to be observed in the geologic record. Some of the events were brought about by changes in the oceans' circulation systems—for example, an influx of colder water to a region that had enjoyed warmer water. Others occurred at times when the oceans, or portions of oceans, became anoxic; that is, they lost oxygen to the point where they couldn't support life. Yet others appear to be associated with times when sea levels dropped drastically and intense volcanism occurred (as the K/T events suggest to many geologists).

If, as it seems, mass extinctions occur on average every thirty million years, then an interesting candidate for a causative agent is large flood basalt deposits in continental regions—the outbreak of new mantle plumes at the Earth's surface, as with the Deccan Traps. Mantle plumes originate at the boundary between the molten core of the Earth and the slightly cooler, overlying mantle, which has solidified. This boundary occurs about halfway down toward the center of the Earth. As heat from the core escapes, it warms the lowermost part of the mantle, which becomes somewhat mushy. The heated and less dense material in the mantle forms a plume, a conduit to carry the heated material up toward the surface. As the material moves up through the resistant mantle, the accumulation of hot, liquid material forms a mushroom-shaped head at the plume's top. When the head breaks through at the surface, there is a violent outpouring of vast amounts of volcanic material over a relatively short period, in geologic terms. Theoretical calculations suggest that the period required for such a cycle in which the lowermost mantle material is heated enough to form one or more mantle plumes and erupt at the Earth's surface is around twenty or thirty million years.

Investigators have sought to correlate such eruptions with extinction events, as well as with cycles of sea-level regressions and transgressions, anoxia, and other phenomena. Any such correlations must be considered speculative, for a number of reasons. They presume that all or most extinctions were caused by similar mechanisms, which may not be the case. Furthermore, just when these extinctions occurred and over what intervals remains obscure. Also, given the paucity of data as we go back in time, some investigators have been selective in the data they have used. And that can, of course, lead to faulty correlations.

Finally, however, and perhaps most important, there is in such studies a presumption that a correlation represents causation, and this can be a serious pitfall in any scientific study. An illustration: if you plot the increase of illumination in city streets in the United States since 1950, you will find that it matches almost exactly the rise in urban crime over the same period—indeed, there is a one-to-one correlation. But it might be a mistake to conclude that street lights improve conditions for criminals, or that turning off the street lights would result in a reduction of urban crime.

PART THREE

MANKIND'S EFFECT ON NATURE

7

Then Along Came Us and We Have Effected Vast Environmental Changes on a Local and Regional Scale

Up to this point we have been looking at the Earth's activities and their effects on climate, on life in general, and more recently on mankind itself. But now, when we look at mankind's effects on the natural environment, we have hit the fast-forward button. We are dealing with a far different time scale. The historical record goes back only around 7,000 years, and even human prehistory—perhaps some 50,000 years—is but an instant in geologic time. Humans have always had some effect on their environment—for example, by the time of the Romans, water pollution was a moderate problem—but only locally, here and there. By 1800, water pollution had become a regional problem in numerous areas around the world. But the most noticeable environmental changes have come about since then, essentially with the beginning of the Industrial Revolution. It is not just the total effects that are of interest; far more important are the rates of change. For if continued indefinitely at increasing rates, the Earth would become largely uninhabitable.

To illustrate the point, human activity, as noted in the last chapter, has led to the extinction of 63 of 4,200 species of mammals, and 88 of 8,500 bird species in the past 400 years. This is an extinction of one and one-half percent of mammals and one percent of birds, relatively small numbers. But if such rates (say, one percent every 400 years) continued through succeeding millennia, the extinction of one hundred percent of all the mammals and birds would occur within 40,000 years. In any geologist's book, that would constitute a new and unprecedentedly rapid mass extinction.

The major extinction events throughout Earth history have occurred over intervals of 100,000 years or so. Furthermore, the anthropogenic extinction rate is almost certainly climbing to more than 1 percent every four centuries. Migratory songbirds, for example, are showing significant declines in population as their wintering grounds in Central and South America are reduced or eliminated by deforestation and as the North American forests (their breeding grounds) are fragmented. If a bird population, such as that of the olive-sided flycatcher, is declining at a rate of 5 percent a year, the species could disappear well within a human lifetime.

Equally astounding is mankind's use of coal and oil—the Earth's reserves of natural hydrocarbon. In geologic terms, it took 500 million years to produce these hydrocarbon deposits and they have been enormously depleted, and at an accelerating rate, in the last 100 years. We are using them up at a rate in excess of one million times their natural rate of production, and this too is an unprecedented change in the geologic environment.

Yet even before the present industrial age, there were environmental problems calling for protective measures. Smoke from wood fires in primitive abodes with poor ventilation has probably always been an annoying but acceptable irritation, but smoke from bituminous (soft) coal, introduced in England in the 1200s as an inexpensive substitute for wood to heat shops and homes, was even more annoying. It combusts very inefficiently, creating great quantities of airborne soot and noxious sulfur dioxide. It was tolerated, however, for simple economic reasons.

But in the 1300s, King Edward I issued an edict prohibiting London merchants from using coal in their furnaces during sessions of Parliament, ordering a return to the practice of burning wood. The penalties were severe: one man, upon being caught with warm coals in his furnace, was put to death. This is the first documented piece of environmental protection legislation. (It was around this time, however, that the Hopi Indians in Arizona found that they could burn coal and did so for making pottery. Legend has it that the smoke smelled so bad that the tribal leaders ordered a kachina—a masked figure representing a spirit-being—to go to all the homes and ban the use of coal indoors.)

In due course, the London merchants prevailed, and the king's edict failed. Episodes of smog continued to plague the city. These were the result of coal smoke and fog, for which London became famous and from which the word "smog" was later coined. Through the years, people continued to object. One such was John Evelyn, who wrote a pamphlet in 1661 with the elaborate title, *Fumifugium, or, the inconvenience of the air, and smoke of London dissipated. Together with some remedies humbly proposed by J. E. Esq., to his Sacred majesty, and to the Parliament now assembled.* The pamphlet begins:

> Sir, it was one day, as I was walking in Your Majesty's Palace at Whitehall (where I have sometimes the honor to refresh myself with the sight of Your Illustrious Presence, which is the joy of Your People's hearts) that a presumptuous smoke issuing from near Northumberland house, and not far from Scotland yard, did so invade the Court, that all the rooms, galleries and places about it, were filled and infested with it, and that to such a degree as men could hardly discern one another from the cloud, and none could support, without manifest inconvenience. It was this alone and the trouble that it needs must give to your Health, which kindled this indignation of mine. Nor must I forget that Illustrious and Divine Princess, Your Majesty's only Sister, the now Duchess of Orleans, who, late being in this city, did in my hearing, complain of the effects of this smoke both in her breast and lungs, whilst she was in Your Majesty's Palace.

Evelyn continued unsuccessfully to seek smoke abatement in London. In 1684, he wrote two notes that are of interest to the scientific historian, one being a simple documentation of the effects of the Little Ice Age, the other being perhaps the first, albeit indirect, notice of a temperature inversion: "I went across the Thames on the ice, now become so thick as to bear not only streets of booths in which they roast meat, and had diverse shops of wares, but coaches, carts, and horses pass over. So I went from Westminster Stairs to Lambeth, and dined with the Archbishop."

On the topic of coal smoke, Evelyn wrote: "London, by reason of the excessive coldness of the air hindering the ascent of the smoke, was so filled with the fuliginous steam of the sea [bituminous] coal, that one could hardly see across the streets, and this filling the lungs with its gross particles, exceedingly obstructed the breast, so as one could hardly breathe."

No amount of complaining would stop the use of coal, and in the 1800s the people of London seemed to take a perverse pride in their smogs, as the use of soft coal increased for both residential and industrial purposes. They began referring to them as "pea soupers," from the yellow color of the smog, or as "London particulars," a word coined by Charles Dickens in his novel *Bleak House*. The heroine describes how a stagecoach driver introduced her to London:

> He was very obliging; and as he handed me into a fly, after superintending the removal of my boxes, I asked him whether there was a great fire anywhere? For the streets were so full of dense brown smoke that scarcely anything was to be seen.
>
> "O dear no, miss," he said. "This is a London Particular."
>
> I had never heard of such a thing.
>
> "A fog, miss," said the young gentleman.

In his story "The Bruce-Partington Plans," Arthur Conan Doyle vividly describes, via Watson, a proper pea souper:

> In the third week of November, in the year 1895, a dense yellow fog settled down upon London. From the Monday to the Thursday I doubt whether it was ever possible from our windows in Baker Street to see the loom of the opposite houses. The first day Holmes had spent in cross indexing his huge book of references. The second and third had been patiently occupied upon a subject which he had recently made his hobby—the music of the Middle Ages. But when, for the fourth time, after pushing back our chairs from breakfast we saw the greasy, heavy brown swirl still drifting past us and condensing in oily drops upon the window panes, my comrade's impatient and active nature could endure this drab existence no longer. He paced restlessly about our sitting room in a fever of suppressed energy, biting his nails, tapping the furniture, and chafing against inaction.
>
> "Nothing of interest in the paper, Watson?" he said.
>
> I was aware that by anything of interest, Holmes meant anything of criminal interest. There was news of a revolution, of a possible war, and of an impending change of government; but these did not come within the horizon of

> my companion. I could see nothing recorded in the shape of crime which was not commonplace and futile. Holmes groaned and resumed his restless meanderings.
>
> "The London criminal certainly is a dull fellow," he said in the querulous voice of a sportsman whose game has failed him. "Look out the window, Watson. See how the figures loom up, are dimly seen, and then blend once more into the cloud bank. The thief or the murderer could roam London on such a day as the tiger does the jungle, unseen until he pounces, and then evident only to his victim."
>
> "There have," said I, "been numerous petty thefts."
>
> Holmes snorted his contempt.
>
> "This great and somber stage is set for something more worthy than that," said he. "It is fortunate for this community that I am not a criminal."
>
> "It is indeed!" said I heartily.

In a speech in 1905, a British physician named Harold Des Voeux introduced the term *smog* and went on to found the Coal Smoke Abatement Society in 1929. The following year in December, thirty people in Belgium's Meuse River Valley died from smog, and thousands more were affected. Seventeen people died from smog in Donora, Pennsylvania, in October 1948. In both cases, smog enveloped the narrow valleys of these two highly industrialized regions and took the lives of elderly people and those with chronic lung and heart diseases.

From such incidents, it gradually came to be appreciated that a major noxious component of smog was sulfur dioxide. In the humid fog, the sulfur dioxide was converted to sulfuric acid, which in turn was adsorbed onto the fine airborne particles of soot. The membranes lining the eyes, nose, and especially the respiratory tract are most vulnerable to these toxic particles, and the results can be bronchitis, bronchial asthma, emphysema, and lung cancer. Beyond human damage, smog blackens buildings with soot and the sulfuric acid erodes both buildings and statuary.

The crisis came in 1952. On Thursday, December 4, a large high pressure system arrived and sat over London, gradually lifting five days after it arrived. All the smoke from millions of residential and industrial chimneys simply accumulated day by day in the cold and motionless foggy air. It was the most intense and long-lasting smog on record, the "killer smog" of

1952. Four thousand people died, chiefly the elderly and those with chronic respiratory problems. Many thousands more may have been permanently affected: an estimated 50,000 to 100,000 were made sick at the time.

The black and yellow smog was so thick that you couldn't see a foot in front of you. People who ventured out wearing white returned with gray clothing; those with colds coughed up black mucus. People got lost in their own neighborhoods. One doctor lost his way trying to visit a patient and wound up on his own doorstep; not to be undone, he enlisted a blind neighbor who led him to the patient's house. Traffic came to a standstill, or barely moved. Cars were abandoned and, in one instance, a policeman on foot, leading a convoy of cars out of London to a nearby suburb, was overcome by smog after walking two miles. Elsewhere, a driver ended up in a cemetery, without knocking down any tombstones and without the slightest idea of how he got there.

After several hundred years of inaction, the killer smog impelled the government to do something about air pollution. In 1956, Parliament enacted the Clean Air Bill, prohibiting the burning of soft coal in London, and the days of pea soupers were soon over.

Across the world, and a little earlier than the killer smog of London, in the early 1940s, the people of Los Angeles began to notice a light blue haze that would occasionally settle over the city, sometimes remaining for days at a time. With it came reduced visibility and varying degrees of eye, nose, and throat irritation. At first authorities decided the irritant was the same as that found in London smogs—sulfur dioxide—and sought to reduce such emissions from various industrial sources. But this produced no noticeable result. Further laboratory investigations showed that petroleum vapors, in combination with photochemical reactions from the Sun's ultraviolet radiation, produced the same compounds observed in the Los Angeles haze. Again, authorities sought to reduce the escape of petroleum vapors from local storage tanks and oil refineries, and again the restrictions didn't help.

A vast network of buses and trolleys that linked the various neighborhoods of greater Los Angeles had begun to be dismantled in the previous decade and the famous freeways were being built. The automobile had become virtually the only means of transportation, and it was finally appreciated that the stinging haze resulted from millions of cars releasing thousands of tons of hydrocarbons into the air daily, thanks to the inefficiency of the engine combustion system.

The Los Angeles pattern has been repeated in many other environmental case histories. It is comforting to be able to point a finger at a specific entity, like an industry, as the polluter. Corrective action can be taken without much harm except to that entity. But it is extremely uncomfortable—and harder to correct—when it turns out, as Pogo said, "we have met the enemy and he is us."

The photochemical haze (sometimes inaccurately referred to as smog) that haunts Los Angeles is made worse by two meteorological phenomena. First, the surrounding mountains essentially prevent air circulation within the Los Angeles basin; the only movement of air is a gentle breeze from the Pacific Ocean. The second phenomenon is the temperature inversion. Generally, temperature decreases upward from the Earth's surface, but not always. Often, warm dry air from the desert to the east flows into the Los Angeles basin at elevations of 1,500 to 3,000 feet, effectively putting a lid over the area and preventing hydrocarbon emissions from escaping upward. Instead, they spread out over the city in a blanket of noxious haze. There are few cities that do not have occasional temperature inversions, and fewer still that do not suffer from photochemical haze. But the Los Angeles basin, which includes all of Orange County and the nondesert portions of three others, still has the worst air pollution in the United States, failing to meet federal air quality standards for four of the six criteria pollutants. It meets the standards for lead and sulfur dioxide, but it is the only area in the nation that fails to meet the standard for nitrogen dioxide; ozone levels sometimes reach three times the health standard, and carbon monoxide and fine particulate matter are often twice the legal limit. Two days out of three, the air is judged to be unhealthy. How unhealthy? It is estimated that if the area met the federal standards, medical costs would be reduced by $9.4 billion a year.

At the same time, nowhere in the United States are more innovative, even extreme, efforts being made to cure the problem. The South Coast Air Quality Management District (AQMD), the air pollution control agency for all of Orange County and portions of Los Angeles, Riverside, and San Bernardino counties, develops and adopts an Air Quality Management Plan every three years. These plans aim to bring about compliance with federal and state clean air standards and include rules to reduce emissions from various sources. According to AQMD, maximum levels of ozone, one of the worst smog problems in the area, have been cut to less than one third of what they were in the 1950s, despite having nearly three times as many people and four times as many vehicles. However, with new, tougher federal air quality

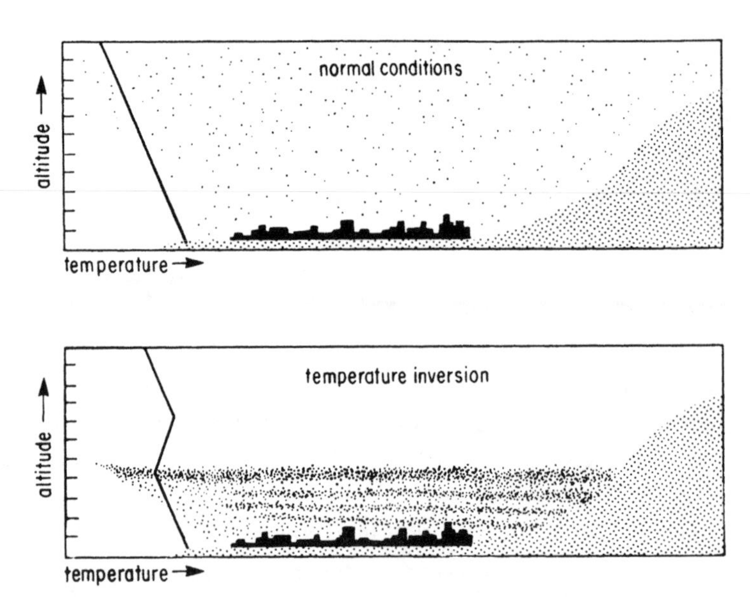

On a clear day in Los Angeles, the temperature decreases with increasing altitude. When the area is capped by a mass of warm, stable air, the temperature increases with altitude for several thousand feet, preventing the dissipation of pollutants from below. Adapted from Wagner, 1971.

standards for ozone and particulates issued in 1997, AQMD estimates it will take until at least 2014 to meet all the new standards.

Photochemical haze appeared first in Los Angeles, ever the trendsetter. Now, of course, it occurs in New York, in Houston (situated on a flat coastal plain with no mountains), in the mile-high city of Denver, where one might expect pure mountain air, in Phoenix, where people used to go to find relief from asthma—the list is long. And it may well be that most cities will have to follow the lead of Los Angeles and, as the English did after the killer smog, get serious.

There has been remarkable progress over the past two decades in cleaning up some forms of air pollution, particularly in the United States, Canada, and much of western Europe. But much of the world's urban population still breathes air that is unhealthy. What usually begins as a local problem in many cases balloons into a regional problem and, sometimes, as we will see later, even a global one.

Much the same situation is found in the waters of the world, particularly freshwater and coastal waters. There is a perception that the deep oceans are the ultimate sump for many of the undesirable waste materials of human activity, but this is an oversimplification. The open ocean is relatively clean; the results of shipping activities and materials precipitating out of the air are found (oil residues, organochlorides, metals, and so forth), but only in low concentrations. Most pollutants are borne by rivers or storm sewers to the coastal waters (bays and estuaries) or to lakes and ponds, and since most contaminants tend to settle out when the water slows down, they simply do not reach the open oceans.

Locally, water pollution is a common phenomenon. Unsightly rafts of algal blooms are common in small ponds near urban areas. Lake beaches are often closed to swimming for health reasons, and freshwater supplies have been reduced. Many coastal bays around the country have been closed to shellfishing because of pollution.

Water pollutants come in two general categories, those that are biodegradable and those that aren't. Lakes, rivers, and coastal regions can assimilate modest amounts of biodegradable agents: human and animal wastes, fertilizers, and detergents. But the exotic components from chemical and manufacturing plants are often persistent, showing no degradation, and some are toxic. First we will look at the history of two large bodies of water that have been overloaded with biodegradable wastes way beyond what could be thought of as modest: the Chesapeake Bay, the largest estuary in the United States, and Lake Erie, the fourth largest of the Great Lakes.

The basic biological cycle is much the same for lakes and estuaries, both of which contain a pool of organic wastes from various sources that is oxidized through bacterial action into inert inorganic compounds. The size of any given organic pool is expressed in terms of how much oxygen is required for the process of oxidization, or its biochemical oxygen demand (BOD).

The nonorganic products that result from oxidation are referred to as nutrients. They include nitrogen found in the form of ammonia, nitrite, and nitrate, and phosphorus, which occurs in the form of phosphate. A third nutrient, silicon in the form of silicates, enters the system through the weathering of soils and rocks. These nutrients, along with carbon dioxide, "feed" the process of photosynthesis, which creates the beginning components of the biological cycle—phytoplankton, the microscopic plants that drift around in the water, diatoms (which need the silicon for their shells), flagellates, and green and blue-green algae among them. Some of these same nutrients will go into producing a complementary pool of rooted aquatic plants.

Under normal circumstances, the phytoplankton pool is consumed by the zooplankton and by herbivorous fish, and these are in turn consumed

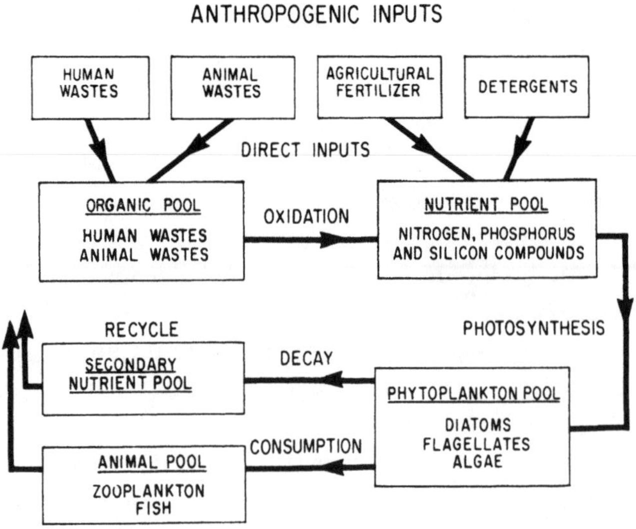

ANTHROPOGENIC INPUTS

HUMAN WASTES
ANIMAL WASTES
AGRICULTURAL FERTILIZER
DETERGENTS
DIRECT INPUTS
ORGANIC POOL
HUMAN WASTES
ANIMAL WASTES
OXIDATION
NUTRIENT POOL
NITROGEN, PHOSPHORUS AND SILICON COMPOUNDS
PHOTOSYNTHESIS
RECYCLE
SECONDARY NUTRIENT POOL
DECAY
PHYTOPLANKTON POOL
DIATOMS
FLAGELLATES
ALGAE
CONSUMPTION
ANIMAL POOL
ZOOPLANKTON
FISH

Basic biological cycle in lakes and estuaries. A critical portion of the cycle is the phytoplankton pool. Increased nutrient availability can lead to unsightly plankton blooms and to anoxic conditions as the available oxygen is used up in the plankton decay.

by carnivorous fish. However, when the nutrient pool is overloaded by various human inputs (nitrogen and phosphorus from fertilizer runoff, for example, or phosphorus from detergents), the phytoplankton pool changes in both quantity and quality. First, with the addition of vast quantities of nutrients, the population of phytoplankton grows enormously, far more than the zooplankton can eat. And because the level of silicon drops compared to the great increases in nitrogen and phosphorus, diatoms become relatively scarce. The diatoms, which are prime-quality food for the zooplankton, are replaced by flagellates, a poorer food source, and by green and blue-green algae, which have no nutritive value whatsoever. Thus, most of the teeming additions to the phytoplankton form a dead end. They themselves must be depleted by decay, not useful consumption. The result of such excessive loading of the aquatic environment, with its excessive BOD, is called *cultural eutrophication.*

One sign of eutrophication is the unsightly algal blooms in affected lakes or bays. The unused planktonic material—chiefly algae and flagellates—eventually sink to the bottom and decay very slowly, over months, using up the available oxygen in the water. This presents no special problem in the winter, when the bottom waters of lakes and estuaries are usually replenished with oxygen from atmospheric influxes at the surface. But in spring the trouble starts. The surface waters of lakes are heated and a thermocline develops between the oxygen-rich surface waters and the waters of the bottom. Literally a heat boundary, the thermocline is a barrier that prevents surface and bottom waters from mixing. A similar process occurs in estuaries, when the spring runoff of fresh riverine water meets the more saline and dense waters of the estuary's bottom. In this case a salt boundary—a halocline—forms, serving as a barrier to mixing. In both cases, however, the planktonic debris keeps on decaying, using up the oxygen in the bottom waters, until a situation of very low oxygen is reached, called hypoxia—or worse, anoxia, which means no oxygen. Creatures that require oxygen to sustain life naturally cannot live in such places; the region is dead. The death of these waters persists through the spring, summer, and fall months until, with the onset of winter, the waters begin to mix again.

This is what has happened to the nation's largest estuary, the Chesapeake Bay, which drains a vast portion of the Middle Atlantic states, including parts of Pennsylvania, Delaware, Maryland, and Virginia, as well as Washington, D.C. The overall bay system includes a number of tributary estuaries, notably the Patuxent, Potomac, Rappahannock, York, and James rivers, and its principal source of fresh water is the Susquehanna River. The amount of municipal and industrial waste, along with agricultural runoff, that pours into the bay is enormous. In 1971, 69 percent of the phosphorus put into the bay was from municipal and industrial waste; 66 percent of the

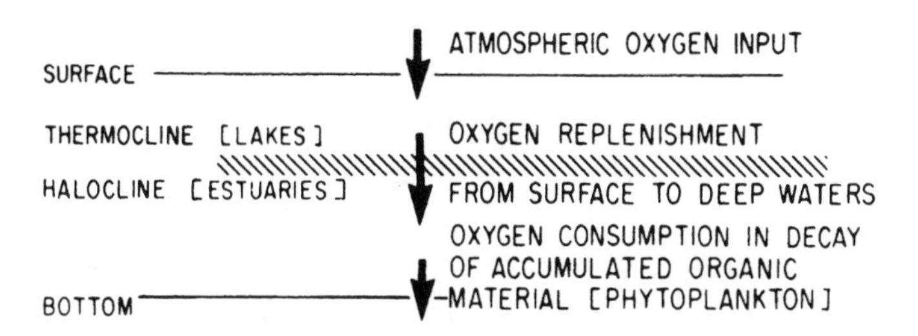

Oxygen dynamics for lakes and estuaries. When the rate of oxygen consumption in the decay of the accumulated phytoplanktonic material on the bottom is greater than the net oxygen replenishment from the surface waters, the oxygen level for the deep waters will decrease and eventually lead to an anoxic condition.

nitrogen came from agricultural runoff. But the process has been going on for a long time.

As far back as the 1930s, it was observed that the deep waters of the midportion of the bay became anoxic for short periods in the summer. From then on the problem grew in extent as well as intensity. During the entire summer of 1950, all the deep waters of the bay channel had dissolved oxygen levels considered hypoxic (which, technically, is zero to two milliliters of oxygen per liter of water). By the summer of 1980, these same waters had become utterly anoxic, beginning in late spring and lasting until early fall, and hypoxic conditions extended over most of the rest of the bay bottom, extending as well up into the Patuxent and Potomac estuaries.

Not surprisingly, there have been major changes in the populations of plankton and fish, and substantial ecological and economic disruption. In the earlier decades of this century, the phytoplankton in the middle and upper regions of the bay were dominated by a prominent spring bloom of diatoms, along with a smaller winter bloom. By the 1980s, these same waters experienced a single large and extended bloom that peaked in summer, dominated by flagellates and green algae. The indirect effects of this change have had a devastating effect on the bay's blue crabs, a prominent, almost legendary, seafood staple for the Chesapeake region: Maryland crab cakes are renowned throughout the country.

By the early 1950s, commercial crabbers reported the death of crabs in the midbay region. Field observations at the time showed that there was greater than 50 percent mortality when oxygen levels reached the hypoxic state. Before 1965, crab fishermen from midbay Tilghman Island found an abundance of crabs in the deeper waters both early in the season (prior to mid-May) and after mid-September. By 1983, there was no more deep-water crabbing. In earlier years, the crabs had hibernated in the mud under the protected deeper waters; by 1983 they were hibernating in shallow waters. Indeed, in 1982, no crabs were caught in waters deeper than twelve feet, and many that were caught were in a poor condition, dying before they could be brought to market.

In the midbay area, "crab wars" have often occurred, in which tens of thousands of crabs crowd into the shallows and even sometimes crawl out onto the land. This bizarre behavior is probably the result of a *seiche*: when the wind blows steadily across the bay, the surface waters tend to pile up on the windward shore (in the manner of a tide). There is a compensatory flow of deeper and poorly oxygenated bottom water toward the lee shore and this obliges the crabs to flee into very shallow depths. Crab wars have also been reported from the Potomac estuary, where, in 1973, all crabs below a depth of eighteen feet died in the hypoxic water. In the Rappahannock estuary, Virginia oystermen have reported "black bottoms" in the late summer of recent years: where dead oysters and other creatures of the bottom are found in black, foul-smelling sediments.

Hypoxia and anoxia are, of course, no better for fin fish than for shellfish. Here too there has been a serious economic decline and a large ecological shift. The Chesapeake Bay had long been a major source of fish like the alewife, American shad, striped bass, and white perch—andromedous fish that travel up the estuary in spring to spawn in shoal waters, with the juveniles heading down the estuary later in the season. From 1960 to 1975, the bay contributed about 3,000 tons a year of striped bass alone to American tables; by 1980, the catch was about 700 tons.

Yet other marine spawners, like spot and croaker, tend to enter the bay in late spring and summer in the form of larvae and juveniles. They are bottom feeders, and anoxia of the bottom waters from May to September restricts their habitat and food sources. Commercial landings of croaker dropped from 20,000 tons annually before 1950 to about 5,000 in succeeding years.

Meanwhile, a less desirable fish has been on the upswing. Landings of menhaden have steadily increased over the past thirty years. These are herbivorous fish that can stand substantial environmental stress (such as low oxygen levels) and they are commonly thought of as "trash fish," used chiefly for fertilizer and animal feed. Menhaden now account for 90 percent of the total fin fish landings in the Chesapeake Bay, a sorry result for a once-rich fishing ground.

The story is much the same in the New York Bight and the waters off Louisiana. In the bays of northern Europe and the Mediterranean, the northern Adriatic, the Caribbean, along the coasts of Central and South America, the Pacific coast of Japan, virtually everywhere people congregate near coasts, algal blooms and sometimes toxic "red tide" phytoplankton blooms are increasingly common annual events in bays, harbors, and offshore. In some cases, the algae count is ten times that of two decades ago.

As gloomy as it all may seem, the situation is by no means hopeless. In both the United States and Sweden, levels of the pesticide DDT in fish and shellfish have decreased significantly, and since a chemical compound called TBT was banned in France and the United Kingdom, their oyster beds are on the way to recovery. And a fairly reassuring tale of how to halt the

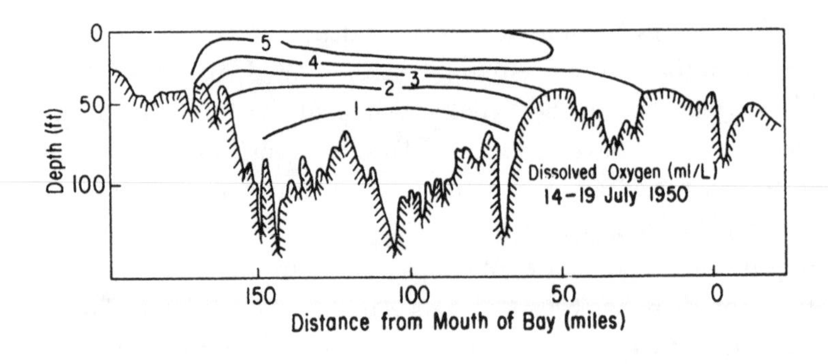

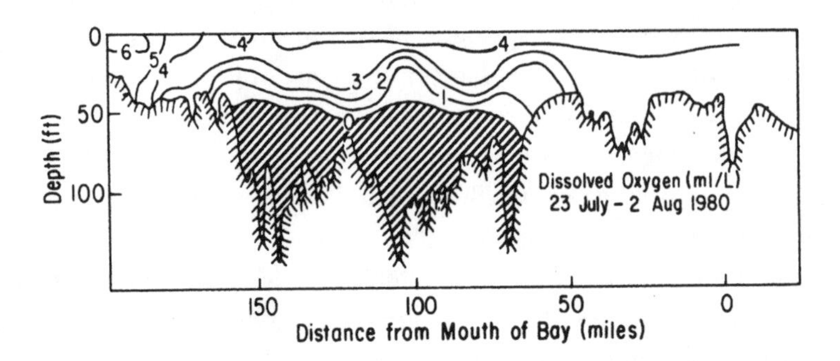

Dissolved oxygen levels in milliliters of oxygen per liter of water along the main channel of Chesapeake Bay extending from the Susquehanna River to the Atlantic Ocean during the summers of 1950 and 1980. Diagonal region is devoid of oxygen. From Officer et al., 1984.

degradation of a large body of water (though not restore it completely) is the effort to save Lake Erie. Indeed, Lake Erie has served as an example of all that can go wrong with an aquatic system.

Lake Erie is the shallowest of the Great Lakes, with an average depth of sixty feet, so it simply doesn't have the assimilative capacity of the others. Furthermore, it has been calculated that it would take two to three years to drain Lake Erie through its natural outlet to Lake Ontario. Not that anyone has suggested doing so; the calculation provides us with the residence time of a particle of water in the lake, the average time that any given particle remains in the lake. Two to three years is a long period of residence, and it means that noxious components—an overabundance of nutrients and toxic wastes—that seep or are dumped into the lake stay in the lake. Beyond that, the shores of Lake Erie are lined with more industries and more cities and towns than any of the other Great Lakes. All of the human and industrial wastes from Buffalo, Erie, Cleveland, Toledo, and even Detroit (via the Detroit River) wind up in the lake, along with fertilizers, herbicides, and pesticides running off from the region's farmlands.

As with the Chesapeake Bay, an early sign that the lake was in trouble came from the fishing industry. In the 1920s, landings of cisco, whitefish, pike, and sturgeon were around fifty million pounds a year. By 1965, landings of these commercially valuable fish had collapsed to a mere 1,000 pounds. The drop in sturgeon could be attributed to overfishing, but not the rest. As in the Chesapeake Bay, there has been a weedy substitution. Fish landings remain near fifty million pounds, but they are what the locals call "rough fish": catfish, carp, and other less-palatable types. These less-valuable fish are warm-water creatures, with lower oxygen level requirements, while cisco, pike, and whitefish are cold-water fishes. They dwell near the bottom, and, as the deep waters of Lake Erie headed for anoxia, these fish headed for oblivion.

By the late 1960s and early 1970s, it was evident that the deep waters of the central basin were going anoxic in the summer. Excessive and obnoxious algal blooms would from time to time cover large portions of the lake's surface. Swimming had to be prohibited in some areas because of untreated sewage and mats of decaying vegetation littering the beaches. Toxic wastes and oil scum haunted the harbors and their entrances. Action was necessary, and over time it was taken. Sewage treatment plants were built; some toxic wastes were curtailed. Most important, there is now a ban on the use of detergents containing phosphorus. From 1940 to 1970, the phosphorus loading of the lake increased threefold, nearly all of it from detergents. Since the ban in 1970, total phosphorus in the lake has decreased by half.

A reasonable question: why concentrate on phosphorus over the other principal nutrient, nitrogen? In photosynthesis, nitrogen and phosphorus combine in an atomic ratio of sixteen to one. Photosynthesis will therefore be limited by whichever of the two nutrients is in the more limited supply according to the combining ratio of sixteen to one. In the oceans, nitrogen is the naturally limiting nutrient, but with the great overloading of nitrogenous waste into Lake Erie, phosphorus was the limiting nutrient. Even major curtailment of the supply of nitrogen would have little to no effect on the algal blooms. But by limiting phosphorus, the photosynthetic process itself is limited, thus cutting back on the plankton crop. So the degradation of Lake Erie has been interrupted by the ban on phosphorus-containing detergents, but the lake is a long way from being in the nearly pristine state it enjoyed before the industrialization of its shores.

Humans have been polluting the world and themselves with toxic materials for millennia. The lead cups the Romans favored are often implicated (however speculatively) in their imperial downfall, and lead poisoning continues to be a problem today, with lead plumbing in older houses, leaded gasoline still required to run old clunkers, and lead paint chipping off old walls to be munched or inhaled by children. Isaac Newton, who was an alchemist as well as the father of modern physics, if not modern science altogether, is thought to have been severely affected later in life by his alchemical dealings with mercury.

It is mercury that brought the severity of toxic emissions into the environment to the attention of a dismayed world not so long ago. In 1953, people living along the shores of the largely inland Shiranui Sea in Japan began to come down with a strange, new, and terrifying disease. Some died, and survivors were left with various combinations of hearing loss, incoherent speech, constriction of the visual field, unsteady gait, and an inability to perform simple functions. By 1956, the disease had reached epidemic proportions, and by January 1975, there were 798 verified cases—all in one area called Minamata Bay. Beginning in 1956, medical researchers at Kumamoto University began investigating, but more than a decade passed before they concluded that the cause of what had come to be known as Minamata disease was the result of mercury. A large chemical factory in Minamata was synthesizing acetaldehyde, a highly reactive industrial chemical that has many uses, including the silvering of mirrors. In the process mercury was used as a catalyst, and the wastes, which included methyl mercury, were discharged directly into Minamata Bay. Neither the bay nor the Shiranui Sea enjoy an appreciable exchange through tides and currents with the Pacific Ocean; the bay thus forms a natural settling basin for waste discharges.

In the course of the investigations, it was discovered that the fish and shellfish in Minamata Bay had mercury concentrations in their tissue ranging from ten to fifty parts per million (ppm). By way of comparison, fish and shellfish from uncontaminated waters elsewhere in Japan had less than one ppm. To compound the problem, fish and shellfish provided an enormous percentage of the diet of this fishing community: a half-pound of fish each day per person in winter, and almost a pound each day in summer—more than twenty times what an average citizen of the United States consumes.

As these matters unfolded in the press, a horrified world looked on. An alarm was ringing in Minamata and it would be heard around the world, again and again, as people began to perceive similar threats to their own environment, their own food supplies, and their own selves from the up-until-then heedless dumping of toxic wastes.

In September 1968, the Japanese government officially declared that the factory's mercury rates were the cause of the disease, and before the year was out, acetaldehyde production was halted. The victims and their families then sought indemnification from the company but were adamantly rejected. Many demonstrated outside the factory and were brutally dispersed by mercenaries hired for that purpose. Finally, in 1973, the district court of Kumamoto assessed a maximum idemnity of $68,000 for fatal or severe cases and a minimum of $60,000 for less severe cases. In its decision, the court said, "In the final analysis... no plant can be permitted to infringe on and run at the sacrifice of the lives and health of the regional residents."

That a halt to the dumping would not immediately clear up the problem was evident from a similar but unrelated outbreak of Minamata disease at Niigata, Japan, in 1965. In all, forty-seven cases were reported from 1965 to 1970. As at Minamata, the disease affected mainly fishermen and their families; their catches were also heavily contaminated with methyl mercury, which derived from wastes from an acetaldehyde factory. In this instance, the factory had ceased operations in 1965.

In 1968, the Japanese government adopted a step-by-step procedure for dealing with mercury contamination in fish. If, in any locality, more than 20 percent of the fish samples exceeded one ppm of total mercury, further surveillance would be called for, which, if necessary, could lead to a ban on fishing imposed by the minister of health and welfare.

Meanwhile, in the 1960s, Swedish ornithologists became concerned about the depletion of the bird population, particularly birds of prey. Investigations finally turned up the cause: seeds treated with methyl mercury compounds as a preservative. As the seed entered the food chain through seed-eating birds and rose up to the avian predators, the mercury compounds became concentrated. Following this, and with the Minamata experience now in mind, Swedish scientists began to look for effects on fish of mercury wastes from paper and pulp mills and chemical plants discharged into the streams. They soon found higher mercury concentrations downstream from such discharges than upstream, but the concentrations were more than ten times less than those at Minamata Bay. No human effects were detected. The Swedish government responded by establishing an absolute standard of one ppm of methyl mercury as the highest permitted for fish on sale.

The following year, 1969, elevated levels of mercury were discovered in predatory freshwater fish in Lake St. Clair and the St. Clair River, two bodies of water that connect Lake Erie and Lake Huron. Researchers found that walleye pike taken from the Canadian waters of Lake St. Clair had average mercury levels of 2.9 ppm, with a maximum of 5.0 ppm. Pike in the river had lower concentrations: an average of 1.6 ppm and a maximum of 2.4 ppm. Before long, similar levels of contamination were found in fish from the American side of the lake. The mercury was traced to waste discharges from a chloralkali plant at Sarnia, Ontario.

This was certainly a matter of concern, though not a cause for panic. The fish-eating habits of the neighboring populace were nothing like those of the people in Minamata Bay, and the mercury concentrations in fish were ten times lower than those of Minamata Bay. No cases of mercury poisoning in humans were reported in the region around Lake St. Clair. Nonetheless, the Ontario provincial government banned commercial fishing in these waters and ordered the plant to put in treatment facilities to curtail mercury discharges. As a matter of national policy, rather than establish overall regulations, the Canadian government chose to treat incidents case by case.

Then the United States ushered in a wholly new era in the history of mercury contamination. On April 2, 1970, the U.S. Food and Drug Administration (FDA) announced that it was "prepared to take legal action to remove from the market any fish found to contain more than 0.5 parts per million of mercury." A kind of reverse bandwagon got rolling. Less than two weeks later, the governor of Ohio banned commercial fishing in the Ohio waters of Lake Erie. Surveys had found walleye pike in the western end of the lake averaging 1.5 ppm, and eastern pike averaging 0.6 ppm—both over the FDA's limit. Similarly, western and eastern yellow perch in Lake Erie were averaging 0.6 and 0.4 ppm, respectively. It soon became clear that mercury pollution was not limited to the Great Lakes; it was a nationwide problem.

In July, the secretary of interior called mercury pollution "an intolerable threat to the health and safety of Americans," and a kind of panic set in. By September, thirty-three states had reported some form of mercury hazard (that is, amounts above 0.5 ppm), and sixteen of those states had imposed sanctions. Losses from both sport and commercial fishing, and fish marketing, began to climb into the millions of dollars, with the Great Lakes fisheries worst hit.

On December 3, 1970, mercury became associated with what almost amounts to a national icon—the little round cans containing tuna. Samples of canned tuna in a local supermarket were found to contain mercury in excess of 0.5 ppm. The FDA confirmed these findings and on December 15, less than two weeks later, the FDA commissioner announced that 23 percent, nearly a quarter, of the 900 *million* cans of tuna packed in the United States in 1970 contained too much mercury. The highest level from 138 samples analyzed was 1.1 ppm; the average was 0.4 ppm. As a precaution (which actually would seem to have been more a shot over the bow of the tuna industry than a consumer protection scheme), the commissioner announced that one million cans were being withdrawn from market.

A little more than a month later, in early February 1971, the FDA reversed its position, stating that the "final statistics showed the problem of mercury to be less serious than had been feared," and that "stocks of the fish presently marketed in the United States are within the guidelines." The final tests in fact showed that only 4 percent of the canned tuna in the United States contained more than the allowed amount. What could have been a disastrous day for the tuna industry—which, before the processed cans of fish are released to the wholesale trade, was worth a quarter of a billion dollars—had been averted.

But swordfish did not fare so well. At the time of the first tuna announcement, early in December 1970, the FDA also stated that swordfish contained 0.9 ppm of mercury. Later that month, it recalled from the market nearly every brand of frozen swordfish, saying that 89 percent of tested samples had mercury levels in excess of the standard.

In 1970, Americans consumed a mere ten million pounds of swordfish—as opposed to almost one-half billion pounds of tuna—and U.S. swordfishermen supplied only ten percent of what Americans consumed. The rest was imported. Indeed, swordfish are ubiquitous and are caught in all of the world's oceans. And, logic suggested, if all this imported swordfish is contaminated at levels in excess of 0.5 ppm, then all the oceans of the globe must be suffering from mercury pollution. Here was reason for real alarm. What with mercury poisoning the American public, and perhaps the entire globe, a great hue and cry over toxic heavy metals of all kinds began, and many instances of dangerous quantities of these substances being dumped into the environment did turn up. Nor was it all industrial in origin. Massive irrigation of parts of California was putting so much selenium in the upper levels of the soil that it was deforming birds.

Meanwhile, some scientists had begun to doubt that the amounts of mercury in most seafood were dangerous or that the world's oceans were

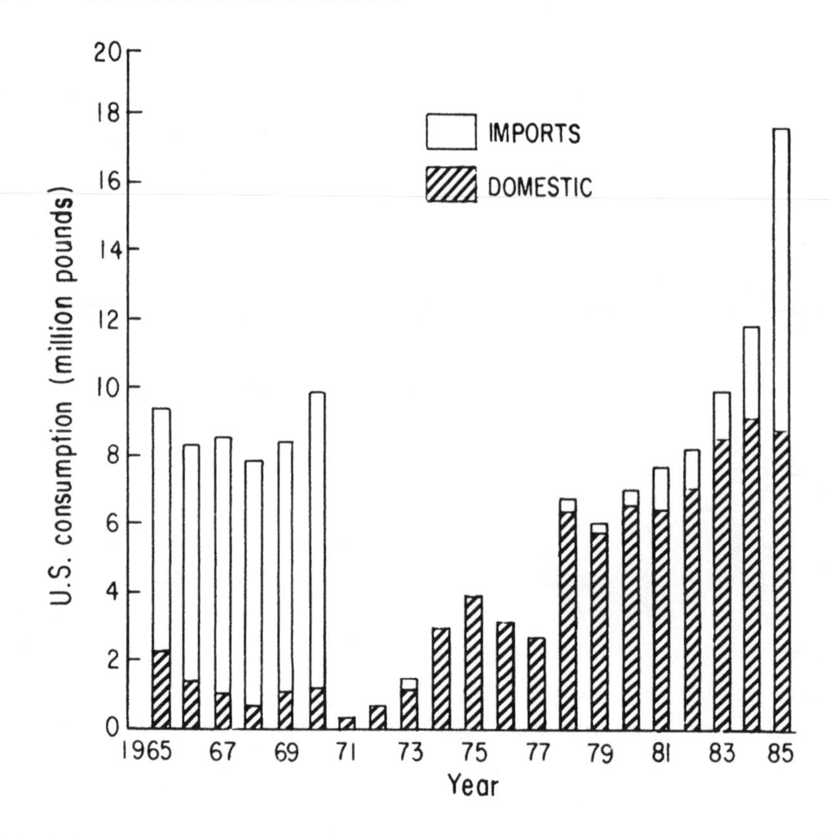

United States consumption of domestic and imported swordfish. From Lipton, 1986.

seriously polluted with mercury. A simple calculation was enough to raise doubts.

The mercury levels of the Minamata fish averaged thirty-two ppm, and the local population ate enormous quantities of fish, as noted earlier. The average level for swordfish, however, was 0.8 ppm. What this means is that if an American ate swordfish for dinner every evening of the year and also had swordfish for lunch in the summer months, he or she would have consumed forty times less mercury than a Minamata citizen. That individual might have some sort of nutritional problem from so restrictive a diet and would have a considerable hole in the wallet, but not mercury poisoning. The problem of mercury poisoning from swordfish did not exist in 1970 and doesn't today.

But it was still legitimate to ask if the world's oceans were becoming polluted with mercury. Again, a calculation suggested no, not seriously.

The mercury content of the oceans is in excess of seventy million tons. From all human-generated sources, an estimated 20,000 tons of mercury enter the sea annually, which is 1/350 of the ocean's mercury reservoir. Most of this mercury is deposited in sediments in shallow coastal waters near the industrial dischargers and doesn't reach the open ocean and its food chain. The amount of mercury entering the saltwater environment, then, from human sources is about the same as that released through the natural processes of rock weathering and volcanism. (It goes without saying that there is no point in heedlessly doubling the natural input, but such simple calculations suggested that the seas were not rapidly filling up with mercury.)

Proof came from museums. If the seas' levels of mercury had changed over the years—meaning because of human pollution—then it would show up in concentrations in the tissue of fish taken years ago and those taken recently. The Smithsonian Institution conveniently had some tuna in its collections that had been caught between 1878 and 1909. Mercury level: 0.38 ppm. For fresh tuna, the level was 0.31. A 1946 swordfish came in at 0.52 ppm, while fresh swordfish were in the range of 0.32 to 1.27. With such figures, it was hard to make a case that mercury was poisoning the oceans, and it was becoming harder still to make a case that swordfish could poison people. In January 1974, the science advisor to the President of the United States had this to say: "Cigarettes, which the Surgeon General and Congress are convinced have health impairing effects on many people, are left on sale. Swordfish, where there is no clear evidence of individual ill effects, are taken off the market."

What, then, was the flap all about? How had what amounted to a panic over mercury arisen? While the United States was in turmoil over seafood and mercury, the same findings were made in Great Britain without any disturbance of the British upper lip. The answer lies in the standard set by the FDA of 0.5 ppm, apparently the result not of medical or scientific analysis but of an error in arithmetic.

Appearing before the United States Senate in May 1970, an FDA official pointed out correctly that the FDA had no direct experience in the area of tolerable mercury levels in fish. He went on to say that the FDA had therefore relied on Swedish investigations. Then he said:

> The Swedish reasoning toward a tolerance from this data was apparently the following:

> The average daily intake of fish was stated to be 200 grams with a mercury content of 50 parts per million dry weight. It was assumed that at 5 parts per million wet weight no poisoning would occur and that a rather low safety factor of 5 would provide an acceptable and safe level of 1 part per million in fish. Thus, the Swedish tolerance of 1 part per million was established. However, it appears that the Swedish work contains a calculation error. A fish is approximately eighty percent water; thus the actual concentration of mercury in the fish, based on wet weight, was 10–20 parts per million. Using the same reasoning as before, the level then becomes about 0.5 parts per million."

Here is what the official was saying: fifty ppm of mercury in dry-weight fish corresponds to five ppm of wet fish—meaning what it weighs when it is caught. This is correct, if the fish is in fact ninety percent water. If a safety factor of five is used, the acceptable level then becomes one ppm in a caught fish, and that was the level adopted in Sweden. But, says the FDA official, a fish is actually eighty percent water, not ninety percent. But this means that ten ppm in wet-weight fish corresponds to the fifty ppm of dry weight. Applying the same safety factor of five, the result is a tolerance level of two ppm and not 0.5 ppm. Had the FDA not established such an unrealistically low standard, there would have been no nationwide scare about tuna and swordfish.

Be that as it may, even as evidence began to build that there was no problem, the FDA continued its ban on foreign imports and interstate sales of swordfish. Imports plummeted from nine million pounds in 1970 to 25,000 in 1975. In 1978, a U.S. district court in Florida found itself in the middle of a legal dispute between the FDA and seafood distributors, finally ruling that a level of 1.0 ppm of mercury posed no reasonable threat to anyone's health. The fishermen and processors had some relief; then matters took a strange turn.

It is often said that human societies and industries take steps that have unknown and unpredictable ecological side effects. It is just as possible to make what might broadly be called ecological judgments that can have unpredictable social effects. Not long after the district court raised the limit of mercury in swordfish to 1.0 ppm, a standard that applied only to imports and interstate sales, some fishermen in Massachusetts (where there is a history of rebellion) noted that the standard did not apply to local fishermen who fished in state waters...or claimed to. So a classic little smuggling operation got underway, with the Massachusetts fishermen heading out for the high seas to buy swordfish from Canadian boats, which were under no swordfish restrictions. Thus, for a few years, most swordfish on sale in New England markets and restaurants entered the country illegally. Eventually the U.S. Customs Service could no longer look the other way, and confiscated a few fishing boats.

In essence, the FDA altered the nature of the swordfishing business in a way it could not have predicted. Until 1970, most swordfish was imported; with the scare, consumption dropped precipitously and imports all but vanished. As the scare began to wear off, swordfish sales began to climb, but all of these gains were local in origin—beyond the jurisdiction of the FDA. An indirect effect of the FDA regulation was the stimulation of a robust domestic swordfishing business. Meanwhile, imports have now returned to take about half the market.

In retrospect, there was no need for the nationwide mercury scare. The FDA's sanction on swordfish was ill-advised and apparently quite hastily determined. Often such regulatory agencies are under the gun to act, but the far-reaching consequences of their actions call for careful consideration, including the participation before the fact, rather than after, by scientists and others whose knowledge might be useful, even crucial.

The instances of mercury poisoning and high levels of mercury in fish were all related to specific industrial or chemical plants. In such cases, proper corrective action is fairly straightforward, if the political will exists, and such actions were taken (though they were a long time in coming at Minamata). The lesson from the American overreaction is obvious, though not necessarily universally understood. But in general, outside of a handful of domestic fishermen, there really were no winners in the mercury story.

In much of western Europe and North America, as well as Japan, the past decades have seen a great rise in industrial and other controls over toxic wastes. Such controls are widely perceived now as "good" for business. But in any highly industrialized state, there are great quantities of these materials produced and in transit at any given time, leading to the occasional spill, or even just sitting in potentially rotting tanks underneath the local filling station. But the modern toxin that incites perhaps the most debate and the most fear is clear, colorless, odorless, utterly invisible, and very hard to understand: radioactivity, particularly radioactivity associated with nuclear power plants.

At the end of 1989, there were 426 nuclear reactors generating electricity, more than 100 of them in the United States, where they produced about a fifth of the nation's electricity. In France, seventy-five percent of the nation's electricity is produced by nuclear reactors. The great advantage of nuclear power has always been seen as using a clean-burning fuel with no emissions of sulfur dioxide and carbon dioxide into the atmosphere, as is the case with fossil fuels. And as a nervous world stares at the greenhouse effect and the results of acid rain (both topics for the next chapter), this advantage looms even greater.

In the years immediately following World War II, nuclear energy was seen to be a panacea to the ever-growing need in the world for electricity. It would be too cheap to meter. It has not yet become anything of the sort, nor will it probably ever be the envisioned panacea. Though it does contribute a significant amount of electricity and will do more so in the future, it is currently stalled in many countries, including the United States. This is partly a result of poor economics and public concerns. It is extremely expensive to build and operate a conventional light-water nuclear power plant; it may have some 40,000 valves alone, compared to 4,000 in a coal plant. This kind of complexity is costly. It also makes people nervous, because so complex a thing is unforgiving of even small errors in construction, operation, and maintenance.

Nonetheless, until recently there had been no calamitous accidents from nuclear power plants. When there has been an accident, such as in 1979 at Three Mile Island in Pennsylvania, the safety systems, cooling systems, and backup systems have worked.

Nuclear power has so far taken a first step, accomplished with the present generation of reactors. These are extraordinarily fuel-inefficient, using only the rare isotope uranium 235 and converting only 0.6 percent of the available uranium atoms. The remaining 99.4 percent are wasted and become part of the spent fuel rods. At this rate, present reactors would use up all the world's uranium ores within a century.

If nuclear power is to have a future, it will depend on a great number of factors, one of which is building more efficient reactors. In fact, second-generation reactors—called fast breeder reactors—are under way. They convert uranium 238 (wasted in present-day reactors) to highly fissionable plutonium, and are about 100 times more efficient. Other designers are concerned with the safety systems, now for the most part a complex of pipes, pumps, chillers, diesel engines, and control systems that bring cooling water to the

Electric Power Generation from Nuclear Reactors for Various Countries as of December 31, 1989

Country	Reactors in Operation	Electricity Generated[a]	Percent of Electrical Power	Reactors under Construction
France	55	52,588	74.6	9
Belgium	7	5,500	60.8	0
West Germany	24	22,716	34.3	1
United Kingdom	39	11,242	21.7	1
United States	110	98,331	19.1	4
Soviet Union	46	34,230	12.3	26
India	7	1,374	1.8	7
Pakistan	1	125	0.2	0
China	0	0	0.0	3
Iran	0	0	0.0	2
Cuba	0	0	0.0	2

[a]Units of electricity generated are in megawatts (MWe).
From Häfele, 1990.

scene in the event of a problem. In the quest for "passive safety systems," one producer has designed a reactor that burns at lower temperatures and sits directly under a vast reservoir of water. All it would take is a change in temperature or pressure, and gravity would bring a flood of coolant to the core. Many of the new reactor designs call for smaller installations, in some cases small enough so that the entire system, including the cooling system, can be enclosed in a single large reactor vessel that could be located in an underground cavity. Other experimental models use liquid metal such as sodium as the cooling medium. With a boiling point of 900 degrees Fahrenheit, sodium can absorb a tremendous amount of heat very quickly. In another design, the fuel is in the form of tiny pellets encased in balls of graphite; in a test at which the reactor was run at full throttle with a simulated total failure of the cooling system, the graphite cases absorbed sufficient heat that the fuel pellets did not melt.

The fear of a nuclear reactor breaking down and exploding like an atomic bomb, releasing great radioactive clouds, is not a reasonable fear, at least according to some experts, particularly in most of the plants now operating, and even less so in those that are slated for the future. The awful exception, however, occurred on April 26, 1986, when the number four reactor at Chernobyl in the Ukraine exploded. A radioactive cloud spread over much of the western portion of the Soviet Union and eastern Europe. The Soviets said at the time that 28 people had been killed (mostly operators and firefighters sacrificially sent in afterwards), 203 people had been diagnosed with radioactive sickness, and some 116,000 people living within thirty kilometers of the accident had been evacuated. But the situation was worse than the Soviets wanted to admit at the time. In 1989, *Pravda* confessed that unsafe levels of contamination extended over an area of 10,000 square kilometers, including large portions of Byelorussia and Russia. Nearly a quarter of a million people lived in the contaminated area—a place where newborn swine and cattle now show genetic defects and high levels of radiation in their tissues. Food for human consumption has to be imported insofar as is economically possible. Children are showing up with swollen thyroid glands, cataracts, and cancers. More than one million people have undergone medical checkups and more than 600,000 have been placed under permanent observation. The region will remain a nuclear wasteland for several hundred years.

The cause of the explosion was a combination of human ignorance and ineptitude on the part of the operating personnel and a design deficiency in the reactor itself. (No reactor in the United States has this design deficiency.) There are, however, sixteen other reactors of the Chernobyl design currently in use; one of them is in Lithuania, where there are no citizens trained to operate a nuclear reactor. It is now being run by a team of Russians in a kind of exile.

One can be assured that if another such accident occurs, it will put nuclear energy on hold for a long time. And there are other concerns, chiefly the ultimate permanent disposal of the radioactive waste products from the fission reaction. The most important are strontium-90 and cesium-137. Both have half-lives of around thirty years—that is, half the radioactive strontium or cesium decays in thirty years, then half the remaining element decays in the next thirty years, and so on. To reduce these elements to safe levels means a period of confinement of a few hundred years. In addition, there are the transuranic elements, like plutonium, which have lower radiation levels but far longer half-lives—some into the thousands of years.

Radiation damage to living things comes particularly from the high energy and penetrating electromagnetic waves called gamma rays. Too much exposure leads promptly to sickness and death. Without noticeable immediate effects, gamma rays can damage enough cells in the body to cause cancer to develop later, or genetic defects in offspring. Further, radioactive strontium and cesium can enter the biological cycle, because strontium can be a substitute for calcium and cesium for phosphorus, both of which are important elements in the food cycle. The mysterious and invisible nature of gamma rays and their delayed effects is more than enough to cause fear, and not entirely irrational fear, among the public.

Yet if reactors can be built for which lethal accidents are not a likely problem (and they evidently can be), the problem remains of how to store the radioactive wastes out of contact with the global fauna and flora for periods of thousands of years. Currently, spent fuel rods are held in water basins at the sites of reactors; there are no long-term disposal sites anywhere in the world. Some have suggested that they can be encased in glass and buried, but the preferred scheme is to bury the wastes in cannisters in crystalline rock formations at great depths, below the strata that groundwater penetrates. Low-level material from the nuclear weapons program is scheduled to be buried 2,000 feet down in an ancient salt deposit near Carlsbad, New Mexico, which has brought forth a loud and prolonged round of the NIMBY syndrome ("not in my backyard"), complicated by the fact that many people in and around Carlsbad are happy at the prospect of the jobs the project engenders (and presumably would for another ten millennia).

The problem seems to be Herculean—as well as Faustian—but it should be noted that the total amount of radioactive waste accumulating each year is about 2,000 tons, which is small compared to the 200,000 tons of other hazardous waste that are generated in the United States each year. In any event, we have indeed produced our Augean stables here on Earth.

We began this chapter by talking about the scales of time and rates of change. In a very short time, less than two centuries, mankind has generated some problems that will challenge our descendants for a period of time that begins to sound almost geologic.

8

...With the Potential for Equally Great Changes on a Global Scale

In the summer of 1988, the ozone was in the wrong place. There was too much of it in American cities and not enough over Antarctica. Unusual weather, in combination with auto and other emissions, had led to the highest ozone concentrations in ten years in many cities. The American public raised a hue and cry for tougher clean air regulations, many of which had been relaxed in the past decade. Fortunately, the 1970 Clean Air Act and its subsequent amendments have produced the desired effects: emissions of some toxic gases have dropped more than 90 percent. Despite the fact that the United States leads the world carbon dioxide emission, recent improvements have disproved the notion that the American public is lethargic on certain environmental matters. Indeed, in both the matter of ozone in urban air and the ozone hole in the stratosphere, the public led its leaders.

About 97 percent of all of the ozone in the atmosphere is found in the upper reaches of the stratosphere, where it absorbs otherwise dangerous ultraviolet radiation from the Sun, thus providing an effective blanket for life. The ozone layer presumably has existed throughout much of geologic time, and the Earth's flora and fauna, including the human species, have evolved without having to contend with extreme doses of ultraviolet radiation. A marked thinning of the layer would cause, at the least, a corresponding increase in skin cancers, along with adverse effects on other forms of life. Throughout the 1970s there were reports of a thinning of the ozone layer, but in 1985 scientists announced that there was an actual void in the layer, a hole that was evidently growing larger at an alarming rate. Once the public began to learn something about ozone and the potential consequences of such a void, reactions were swift.

Ozone (O_3) is a colorless gas, a molecule of which contains three oxygen atoms. The intense sunlight in the high elevations of the stratosphere produces ozone naturally by breaking a normal oxygen molecule (O_2) down into two highly reactive oxygen atoms (O). Then each oxygen atom quickly combines with a normal oxygen molecule to form ozone. In turn, an ozone molecule absorbs ultraviolet radiation and in the process is changed back into an oxygen molecule and an oxygen atom. A balance exists between production of ozone and its destruction, leading to an equilibrium concentration of ozone in the stratosphere. But add an alien substance that enhances the destruction of ozone and the equilibrium is gone. And, of course, this is just what happened once humans invented the useful chlorofluorocarbons, or CFCs. They have been in use for the past sixty years as coolants in refrigerators and air conditioners, and as propellants in spray cans. CFCs are nonflammable, nontoxic, and stable, meaning that they cannot combine chemically with other substances. Stability is one of the features that recommends them. But once they are let loose in the air, they diffuse slowly upward into the stratosphere, where they are attacked by the intense ultraviolet radiation and break up into their primary components, one of which is chlorine.

In the ozone layer, chlorine acts as a catalyst, enhancing a reaction without itself being changed. And the reaction these interlopers enhance is the destruction of ozone. It works like this: a chlorine atom (Cl) combines with an ozone molecule (O_3), and the result is a molecule of chlorine monoxide (ClO) and a normal oxygen molecule (O_2). But there are also oxygen atoms (O) in the neighborhood, and the chlorine monoxide molecules (ClO) react with them, producing a normal oxygen molecule (O_2) and a free-ranging chlorine atom that can head off and react again with ozone, destroying it. A relatively small amount of chlorine can do a lot of damage.

For example, the atmospheric levels of one of the more prevalent CFC compounds, CFC-12, were measured over Barbados. The results showed an increase from 260 parts per trillion (ppt) in 1978 to 420 ppt in 1987, almost doubling in less than a decade. Still, parts per trillion is an unimaginably minute measure: the CFC-12 concentrations are equivalent to less than one person out of the entire population of the world. It would be reasonable to ask if such tiny amounts of even a virulent catalyst could have a recognizable effect on the ozone layer. But measurements over Antarctica and to a lesser degree over the Arctic suggest that this is the case.

It is sunlight that powers the creation (and destruction) of ozone in the delicate equilibrium we mentioned. But during the winter months over Antarctica, there is no sunlight, so there should be no ozone layer—and there wouldn't be except that stratospheric winds carry in ozone-rich air produced in the temperate and tropic latitudes. Yet in spite of this dynamic process, the ozone layer over Antarctica in winter has decreased by about half since 1970. In some years, the hole in the layer has reached an area the size of the continental United States, but it seems to fluctuate in size. For example, it was large in 1987, less so the next year, and then, in 1989, back to its 1987 size. This may have something to do with temperature.

During the long, dark Antarctic winter, polar air forms a great vortex so cold that icy clouds form in the stratosphere at altitudes of about ten miles. There is reason to believe that the ice in the clouds reacts with CFCs, breaking them into various constituents and producing free-ranging chlorine molecules. Then, when the sunlight returns in August, it breaks them down into the lethal catalytic chlorine atoms that destroy ozone. As the air warms up, the clouds disappear, and the chlorine atoms bind back into molecular forms, halting the ozone depletion process. It seems that the colder the temperature of the vortex, the more severe the problem. If this sequence of events is correct, then the ozone hole over Antarctica probably will not get any bigger than it was in 1989, because the vortex of polar air is hemmed in by other weather systems. Nonetheless, it is not strictly an Antarctic problem. Ozone-depleted air has been found to leak out of the ozone hole, producing depletions of 10 percent over New Zealand and Australia.

Most scientists studying this matter believe that some such system is at work and that the CFCs are the culprit, though there are nagging doubts that some or all of the decrease may be connected to some dynamic changes in the stratospheric wind system. The measurements have been made over only a few years, after all, and a good deal of uncertainty remains about the physics and chemistry of stratospheric process and events.

Nagging doubts among scientists or no, the public reacted by deciding it could do without spray cans that use CFCs as propellants. Recognizing the futility of producing an item that people won't buy, industry was quick to follow, developing substitute propellants that they could advertise as new and improved and environmentally safe. In 1987, the U.S. government banned the use of CFCs in spray cans and this in turn was followed by an international agreement to stabilize and ultimately eliminate the production of CFCs globally. When this occurs, it will be none too soon, since CFCs are also an extremely potent greenhouse gas.

The lessons to be gained are threefold: (1) The public can act, dragging industry and governments along; (2) scientists remain ignorant, nevertheless, about many matters that are of major importance, such as the physics and chemistry of the stratosphere, when we try to understand the nature of anthropogenic effects on the planet; and (3) it sometimes doesn't take that much—in this case the production of relatively innocuous chemicals in relatively small quantities—to have a potentially global impact.

It can be argued that the CFC problem was unpredictable, that it sneaked up on the world, and it can also be argued that it may have been nipped in the bud before it actually became a global calamity. A number of anthropogenic assaults, however, are also not yet strictly global problems, but they were not only predictable, but predicted. Acid rain is one of these. Forestry scholars at Yale University and Dartmouth College were concerned as long ago as the early 1960s by the effects on New England forests from what they were certain was acidic rain produced from the burning of fossil fuels, specifically sulfur dioxide. At the same time, fishermen in Scandinavia, then in Scotland and North America, began to notice a striking decline in both the quantity and quality of fish that were caught in some remote lakes. It was not long before the same lakes were shown to have become more acidic.

The measure of the acidity (or conversely, alkalinity) of a given solution is the pH number. The pH scale ranges from 0 to 14, with a neutral solution having a value of 7. Anything greater than 7 is alkaline (or basic); anything less than 7 is acidic. The pH scale is logarithmic (rather like the Richter scale for earthquakes), so a solution with a pH of 4 is ten times more acidic than one that is pH 5, and 100 times more acidic than one that is pH 6.

By way of benchmarks, lemon juice has a pH ranging from 2.2 to 2.4. Vinegar is 2.4 to 3.4, and grape juice is 3.5 to 4.5. Normal rain, which is slightly acidic with a pH of 5.6, is at the same time ten times less so than grape juice. The moderate acidity of normal rain is a result of the combination of ambient carbon dioxide in the atmosphere with water vapor, forming minute droplets of carbonic acid, a relatively mild acid. Any rain with a pH of less than 5.6 is considered acid rain. Today the acid rain that falls on the northeastern part of the United States has pH values ranging from 4.3 to 4.5. In the summer clouds overhead, the lower portions have values around 3.6, sometimes reaching the acidity of vinegar.

This is largely the result of droplets of sulfuric acid formed when sulfur dioxide combines with water vapor in the atmosphere, and more than half the sulfur dioxide released into the atmosphere by human activity comes from burning coal. In the United States, 85 percent of the coal burned is to produce electricity. United States coal consumption doubled between 1937 and 1988, and the great proportion of coal-fired power plants exist in the industrialized Midwest. Early on, high smokestacks were installed to eliminate locally the foul emissions of smoke and soot, and of course these products were dispersed downwind. The sulfur dioxide too was swept several hundred miles downwind to be deposited as acid rain. This has pitted New England against the Midwest, Canada against the United States, with similar problems in Europe and in almost every industrialized area that has neighbors downwind. Acid rain is not only a problem of industrialization. Each year African farmers and herdsmen burn vast portions of the savannas—some 75 percent of such lands—to clear them of pests, insects, and dead grass. Scientists have recently suggested that acids released in the fires are swept by the northeast winds over the forests of central Africa, contributing significantly to the acidity of the humid air, with its pH ranging between 4.4 and 4.6.

Just as hypoxia and anoxia cause ecological rearrangements in a lake or estuary, so does a change in acidity. The fish at the top of the food chain (called high-trophic fish, typically considered the most desirable) are the most affected, while low-trophic "trash" fish can handle considerably lower pH levels. One Canadian lake was intensively studied over a period from 1961 to 1973, during which the pH level changed drastically. The first species to

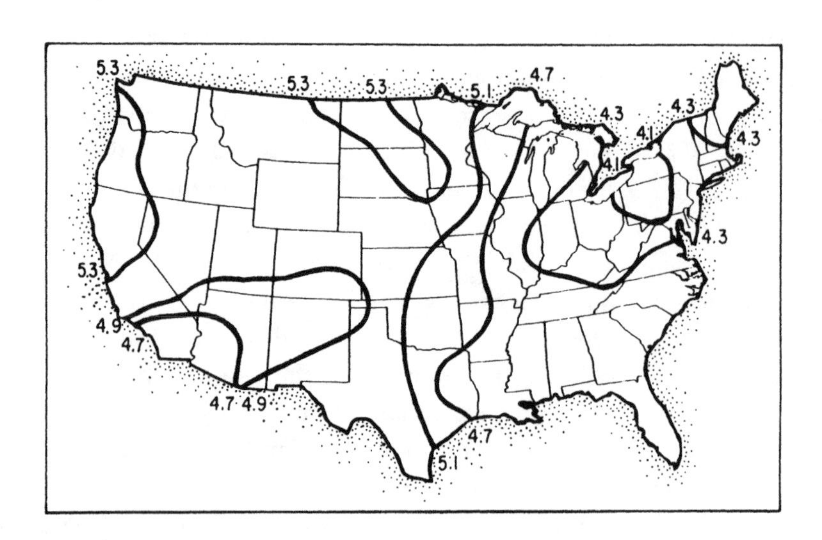

Acid rain precipitation in the United States. Contour lines chart the average pH of the precipitation. From Mohnen, 1988.

disappear was smallmouth bass, followed by walleye, turbot, and lake trout at pH levels of 5.8 to 5.2. Then, as the pH dropped to 4.7, northern pike, white sucker, brown bullhead, pumpkinseed sunfish, and rock bass all disappeared. Yellow perch and lake herring were still spawning in the lake as of the end of the study in 1973 with the pH at 4.7.

The acidification of lakes was one thing; in the early 1980s, trees in the Black Forest of Germany and other European forests began to die at an alarming rate. By 1985, it appeared that more than half the trees in the Black Forest were damaged. People suggested climate changes, drought, disease, insects... but none of these could be shown to be the culprits. Similar problems were spotted in the forests of eastern North America, especially those at high elevations. Indeed, in the 1970s and 1980s, more than half the red spruce trees in New York's Adirondacks, the Green Mountains of Vermont, and the White Mountains of New Hampshire have died. Acid precipitation is taken up in the roots of the trees and tends to reduce their tolerance for cold. Canadian studies showed that the most sensitive trees include Douglas fir, eastern white pine, white ash, and white birch. Those that show immediate effects include balsam fir, basswood, red pine, and white elm. Those that are tolerant of acid rain include balsam poplar, red oak, and sugar maple. The conclusion was, and is, inescapable that acid rain has had devastating effects on lake fish and forests, here and abroad.

Meanwhile, Canadian environmentalists and other groups were pointing the finger at the Midwestern power plants and seeking redress. In the late 1970s, the U.S. Congress decided to institute a program to evaluate the situation, the National Acid Precipitation Action Program (NAPAP). Its mandate was to study the science of acid rain and come up with recommendations for an action program—for example, answers to such questions as what environmental benefits would accrue from a reduction of, say, 20 percent of sulfur dioxide emissions.

NAPAP got started in 1980. After several changes in leadership and direction, it issued its report in 1990, all 6,000 pages. A photograph shows a staff member standing beside the report, which is the same height as she is. It cost $500 million to produce, the work chiefly of the Environmental Protection Agency, the Departments of Interior, Energy, and Agriculture, and the National Oceanic and Atmospheric Administration. As can be imagined, a vast amount of data were collected. Yet those data contributed little to the *science* of acid rain. Data collection per se cannot be equated with science: in a scientific study, the results from one investigation may lead to an entirely different approach toward understanding the problem, calling for new and different data. This did not occur in the years of NAPAP. The masses of data collected may or may not eventually be of value to science.

Nor did NAPAP produce any suggestions for action, in spite of a gigantic computer program devised for that purpose. Any useful computer model has to rest on those scientific understandings that inform the program. Otherwise it's the old story: "garbage in, garbage out." What did come out were four major conclusions: (1) acid rain had adversely affected aquatic life in about 10 percent of eastern lakes and streams; (2) it had contributed to the decline of red spruce at high elevations; (3) it had contributed to the erosion of buildings and materials; and (4) along with related pollutants and especially fine sulfate particles, it had reduced visibility throughout the Northeast and in parts of the West. Every one of these conclusions could have been stated without benefit of this half-billion-dollar study.

Nevertheless, progress was being made even while NAPAP dithered. Impelled by public concern and legislation from Congress, many of the utility companies in the Midwest substituted low-sulfur fuels for high-sulfur coal. (This had the ramifying effect of putting high-sulfur coal miners like those in West Virginia out of work and also increasing strip mining in the West, where the coal is of a lower sulfur content.) Scrubbers have been installed on many stacks. These consist of a limestone slurry or other absorbent, which reacts with sulfur dioxide to form a compound that can be removed as solid waste. A standard scrubber can remove from 50 to 90 percent of sulfur dioxide emissions, but with some cost in plant efficiency. These costs are ultimately passed along to the consumer. In any event, as a result of scrubbers and low-sulfur fuels, sulfur dioxide emissions have been reduced by a third in the past fifteen years, a substantial (and commendable) first step.

Many such environmental problems, though nearly global in extent, are not totally intractable. A search for substitute materials and a determined public can perhaps halt ozone depletion. More efficient controls can prevent acid rain. But there are larger problems that do not admit of anything like a simple technological fix or a bit of international political cooperation between neighbors. They are the truly global challenges and they all stem from precisely one source—the proliferation of human beings and their demands for resources.

In 1800, there were one billion people on the planet; there are 6.77 billion in 2008. Pessimists among demographers project that the next century will end with ten to sixteen billion; optimists project that it will stabilize at five to six billion. In North America, the population has increased from a mere 26 million in 1800 to an estimated 514 million in 2008. The growth curve leveled off during the Depression years of the 1930s, rose sharply in baby boom times of the 1940s and 1950s, and has since leveled out again, possibly permanently. Similarly, the population of western Europe appears to be leveling off to essentially a replacement rate. The gloom and doom arises largely from the developing world, which is well on its way to exonerating the prophet of gloom and doom in population analysis, Thomas Robert Malthus, a nineteenth-century Anglican clergyman who studied mathematics at Cambridge University. While it is a truism that population must always be kept to a level that can be supported by the food resources of the land, Malthus put this concept into quantitative terms in a series of treatises, beginning in 1798 with *An essay on the principle of population as it affects the future improvement of society*.

Malthus started with some simple mathematical considerations, comparing the hare of population increase to the tortoise of subsistence increase, and concluding that the race would always go to the fleetest—population increase. For example, assume that population in a given region doubles every twenty-five years, producing a sequence of 1, 2, 4, 8, 16, 32, 64, 128, 256, 512, 1,024.... If the region were a small state like New Hampshire, with a 1980 population of about one million, the state's population would reach one billion in the next two and a half centuries. The notion of a billion people crammed into tiny New Hampshire is inane, but it serves to illustrate Malthus's concept that population, like compound interest, increases in a geometric or exponential fashion and can easily get out of hand. He framed what came to be called Malthus's dismal theorem:

1. Population is necessarily limited by the means of subsistence.
2. Population invariably increases where the means of subsistence increase, unless prevented by some very powerful and obvious checks.
3. These checks, and the checks which repress the superior power of population and keep its effects on a level with the means of subsistence, are all resolvable into moral restraint, vice, and misery.

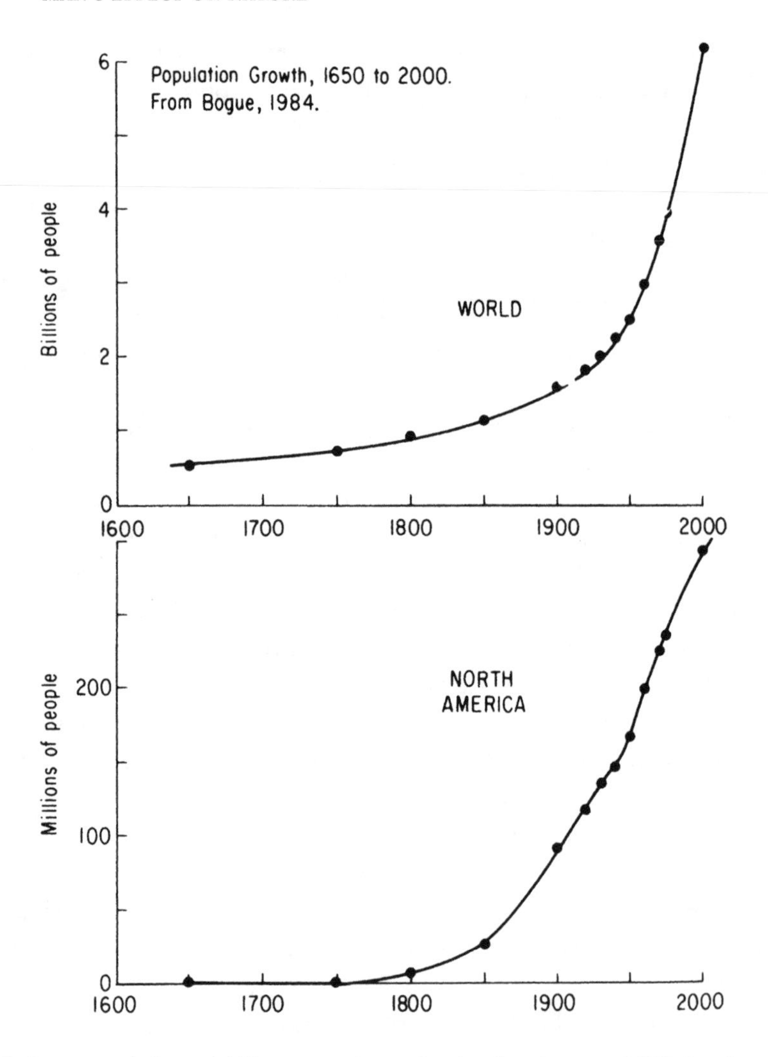

Population growth from 1650 to an estimated value for the year 2000 for the world and for North America. From Bogue, 1984.

Malthus's ideas were controversial from the start and for the most part unpopular. He was looked upon wrongly as an advocate of war, pestilence, and other forms of misery, along with vice, as checks on population. His later opinions, such as opposition to the English Poor Laws that established the right of the poor to relief, led people to see him as a toady of the rich. Furthermore, it turns out his beliefs about population growth did not apply to the Europe of his day and were widely dismissed. Two major factors permitted the tortoise of subsistence to outstrip the hare of population: the Industrial Revolution, with its great increases in both industrial and agricultural production, and the opening up of the vast spaces of the Western Hemisphere.

A Malthusian exception took place in Ireland. With the introduction of the potato from the New World, the Irish population soared to more than eight million, according to the census of 1841. But four years later, with the arrival of potato blight, the crops failed, and again the following year (1846). An estimated one million people died of starvation; many others emigrated to North America. Eventually, the population stabilized at four million, or half the preblight figure, a continuation of emigration and Malthusiasn "moral restraint" in the form of late marriages.

Today, Malthusian vice and misery appear to be operating in the underdeveloped and overpopulated regions of Asia, Africa, and South America, where urban populations are growing exponentially, along with all of the problems associated with urban crowding, poverty, and environmental degradation. The vagaries of the monsoonal excursion to the north in Africa's Sahel have led to hundreds of thousands of deaths *annually* from starvation. The cyclones and typhoons that strike the densely populated lowlands of the Indus delta region regularly take catastrophic numbers of lives. Wars, both large and small, seem constant, and AIDS (and other diseases like tuberculosis) is on the march, devastating Africa in a manner reminiscent of the bubonic plagues of earlier centuries. The limits of subsistence would seem to have already been reached in such regions, yet in many of these countries the population growth rate remains high.

We in the United States may well be reaching certain kinds of limits to subsistence growth. Many years of drought in California have served to highlight that state's always profligate use of water. People in the San Francisco area, which enjoys a relatively moist climate compared to southern California, are becoming inured to quite draconian restrictions on the use of water. Now some of the cities surrounding San Francisco have restricted the water allowance for households. By and large, the populace has accepted these restrictions, which limit water use for gardening, swimming pools, and car washing. In the arid south, cities absorb water like vast sponges, always in competition for this scarce and imported resource with the farmers of the Imperial Valley. There is even talk of piping water in from Alaska. Indeed, throughout the West, fresh water is the fundamental issue between states, between locales and cities, between industries, even between forms of agriculture. The rights to more than 100 percent of the annual waters of the Colorado River have been sold; the river itself, even as it courses through the Grand Canyon, is now little more than an artifact. At the other end of the water use cycle, one finds similar restrictions. The municipal sewerage system in the Boston metropolitan area has simply reached its limits: it has been mandated that no new residences or other buildings may be connected to the system. That means no new construction, which, until the system is somehow expanded, means no growth. Even in the spacious United States, we seem to be reaping the harvest of an extremely rapid growth of population in the last two centuries.

The population growth rate obviously depends on the difference between the birth and death rates, but what is not so obvious is that the controlling factor has been the death rate. With more plentiful and varied food, with many infectious diseases under control, and with improved public health measures decreasing infant and child mortality, the average life expectancy in the industrial nations has increased from twenty-five years in 1800 to about seventy-eight today. Continued dramatic improvements in life expectancy are not considered likely, and that in itself suggests that the large increase in population over the past 200 years may be a one-time effect. Meanwhile, birth rates have actually declined, though not far enough to offset the increase in longevity. Birth rates typically do not adjust themselves as quickly as death rates, which are, of course, immediately affected by an increase in life span. In earlier times, large families were a necessity (to offset high death rates and to create a large-enough local labor force) and therefore a civic virtue. Such attitudes take time to change, but the staggeringly high costs of child rearing and education are among the factors changing old ideas.

To produce such a change in attitude in the developing world, where the population problem is obviously the most intractable, seems almost impossible. With relatively poor public health systems, there is high infant mortality—the specter that drives families to have many children. There is a great deal of economic and environmental stress on people in poverty, and it has recently been shown that young women under stress reach puberty and thus child-bearing age earlier than normal. It is estimated that there will be by the year 2000 more than 1.8 billion women of child-bearing age in the less-developed regions of the world (as opposed to about 290 million in the developed regions). The trend seems overwhelming, but recently another trend has come to light, suggesting what almost amounts to a quick fix: education, and particularly education of women. In most developing nations, more females die young than males; they get less medical care and less nutrition. Maternal mortality in developed countries rarely exceeds 10 per 100,000 live births. In the developing countries, the rate is between 400 and 1,000. In the developing world, on average, female literacy is three-quarters that of the male; in many of these countries it is half. Ideally it takes only one generation to bring female education levels up to current male rates, or beyond. And there seems to be almost a natural law involved. According to the United Nations' *Human Development Report 1990*, higher female literacy seems to lead very quickly to "lower infant mortality, better family nutrition, reduced fertility and lower population growth rates."

In Bangladesh, for example, child mortality was five times higher among children of mothers with no education than for those with merely seven years of schooling. Colombian women with the highest levels of education had four fewer children than mothers with only primary education. Still unconvinced? In five countries (Yemen Arab Republic, Afghanistan, Mali, Sudan, and Pakistan), female literacy was found to be only 20 percent or less in 1985...and the population growth rate was 3 percent or more. In five other countries (Dominican Republic, Jamaica, Sri Lanka, Colombia, and Thailand), female literacy was 80 percent and the population grew at 2 percent or less.

The lesson seems clear enough. Before investing in dams, or roads, or elaborate technical apparatuses of either peace or war, developing countries can help themselves most (and almost overnight) by investing in education, particularly that of women. And there is probably no constituency in the United States that would object to the federal government and various international development agencies and banks tying their aid and trade policies to female education around the globe. The biggest investment seems to be getting little girls into the classroom and keeping them there for as many years as possible. Such may be the only way to replace the Malthusian option of misery and vice with moral restraint.

Meanwhile, in Malthus's theorem, if you substitute the phrase "nonrenewable natural resources" for "subsistence," the consequences are even more dire. After all, food is replenished seasonally, but such things as fossil fuels and minerals are not. Once they have been used up, that is that; they simply are no longer available. And graphs of the consumption of energy and mineral resources look very much like the exponential growth curves of population.

We discriminate the major eras of humankind's time on the planet by what we have extracted from the Earth beneath our feet. The Stone Age extended from prehistoric times to 3000 B.C. (and beyond, locally). Next was the Bronze Age (bronze being an alloy of copper and tin) from 3000 to 1000 B.C., then the Iron Age from 1000 B.C. to A.D. 100 or the present, depending on one's choice. Or one could say that we now live in the Oil Age, which began late in the last century during the "romantic" period of the extraction industries when engineers and entrepreneurs eagerly roamed the Earth's surface, looking for mineral and oil riches.

The heady days of mineral and petroleum exploration are over, times when the rugged individualists prospected, made fortunes, lost them, then moved on, opening up remote areas to settlement. Whether the mythology corresponds with the reality of those times or not, the reality today is sobering. You need to drill far deeper, or offshore, for oil, and only huge corporations or conglomerates of small firms can play the game. Most of the major, near-surface, easily mineable ore deposits of minerals have been found and exploited, and production comes from increasingly great depths and from lower-grade deposits. We are squeezing the last juices out of the fruit and, as a result, they are getting more expensive.

In 1850, wood supplied 90 percent of all the energy used in a young United States—a renewable resource, by the way. By 1910, coal supplied 75 percent of American energy, and by 1970, oil and gas provided 75 percent. Meanwhile, the amount of energy used has risen exponentially, as is clear from the accompanying graph. What it does not make clear is that while the U.S. population will have increased roughly sixteen times from 1850 to 2050, the increase in energy consumption is three times that: it will have grown forty-eight times over the same period. Worldwide, energy consumption appears to be rising at a rate of about 3 percent annually, the major proportion of it fossil fuel, and just at the point in geologic history when the by-products of burning fossil fuels are especially unwanted. We will explore some of the ramifications of this—notably the greenhouse effect—later on. For now it is worth asking how long this profligate mining of fossil fuels and other nonrenewable resources can continue. A look at the accompanying table on crude oil, the major energy source of the United States and the rest of the industrialized world, is far from reassuring. Production worldwide in 1988 was 21.1 billion barrels and the ultimate reserve is 2,074 billion barrels. Simple arithmetic shows that, at the 1988 production rate, all of the reserves—all of the oil in the world—will be gone in 100 years. It will

World Crude Oil Production, Identified Reserves, and Ultimate Predicted Resources (billions of barrels)

Region	Cumulative Production through 1988	Production in 1988	Identified Reserves	Ultimate Resources
North America	182.8	4.4	83.0	387
South America	57.9	1.4	43.8	142
Western Europe	15.7	1.4	26.9	69
Eastern Europe	6.8	0.1	2.0	11
Soviet Union	103.6	4.5	80.0	285
Africa	46.4	2.0	58.7	153
Middle East	160.2	5.1	584.8	867
Asia/Oceania	36.8	2.2	42.8	160
World	610.2	21.1	922.0	2,074

From Masters et al., 1990.

have taken only two centuries to exploit every bit of what it took the Earth a half-billion years to produce. And in fact it will probably be gone sooner, since the rate of production will almost surely increase over the next decade or two. Another uncomfortable and well-known fact is plainly shown in the graph: if we rely on oil, we are going to rely on the political and international responsibility of the Middle East, the most volatile place on Earth. For there, in the crucible of most of the world's major religions, are also found 63 percent of the identified world reserves of oil and 42 percent of the ultimate predicted reserves. Indeed, by 2008, even economists were in agreement with geologists that peak oil would occur by 2010. Peak oil means that petroleum drilling and production has reached its upper limit, and we will steadily be obtaining less and less petroleum. From here it will be, in other words, all downhill, except for oil prices and a rising clamor to find substitutes, especially for transportation but later for all energy uses. The corresponding figures for natural gas give little more encouragement; at current rates of use, the world supply will last 160 years. For coal, supplies will continue for a few hundred years, perhaps as long as 500. Of the three, coal is the dirtiest, and the more that is left in the ground unburned, the better.

It is far more difficult to make accurate predictions of the reserves of minerals. Any given deposit can vary widely in its elemental or mineral concentrations from one location to another. Veins playing out are a common frustration of the miner. Specifically, we have imprecise knowledge at best about how deep ore bodies extend or where their irregular boundaries with the host rock run. A falsely reassuring rule of thumb says that the ore reserves

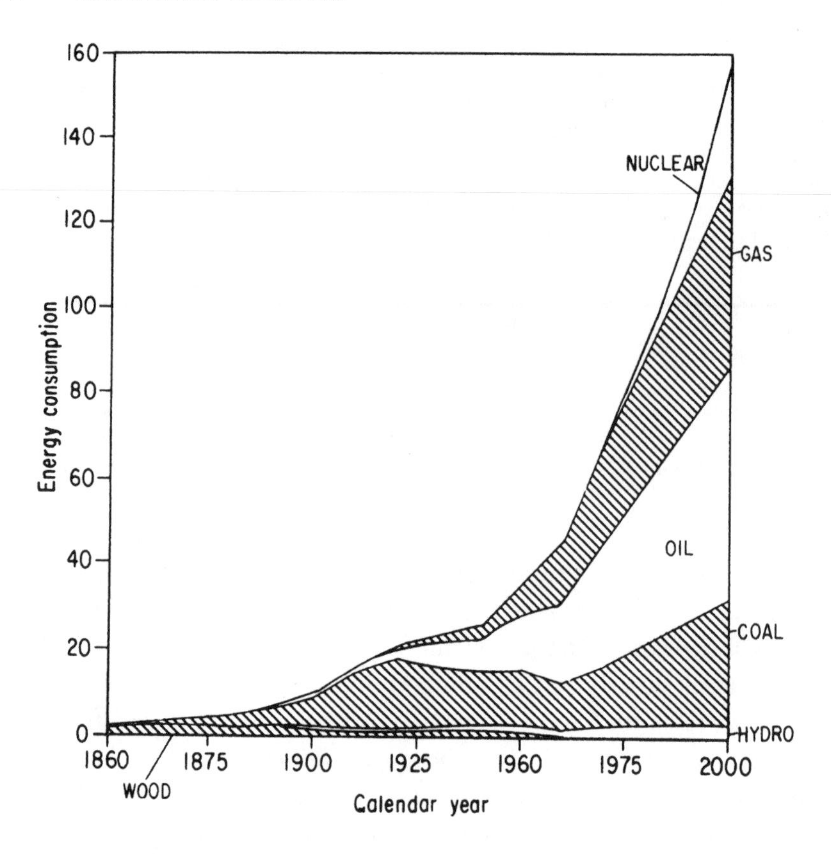

United States energy consumption from 1850 to its projected level in 2000. Energy scale is in 1015 BTU (British thermal units) per year. From Starr, 1971.

increase geometrically as the grade of the ore decreases arithmetically. This suggests that vast if not unlimited supplies of a given ore exist, the only real problem being that it gets more expensive to tap those supplies.

In fact, the rule does seem to apply to some ores, such as copper, but not to most. For example, the ores for metals like mercury, tin, tungsten, and tantalum are concentrated in a few small deposits where the demarcation between ore and host rock is sharp. There is no gradation of increasingly low concentration as you move away from the main ore body. These and other similar deposits of critical mineral deposits will be exhausted within the next fifty years. At the other extreme are huge low-grade deposits of iron and aluminum that are virtually ubiquitous. In between are relatively common minerals such as copper, cobalt, nickel, and vanadium, along with lead and zinc.

The solution to the dwindling of mineral supplies is obvious: recycling. Both industrial and residential recycling is beginning to catch on, and while it is a good and noble thing for a householder to recycle aluminum cans and newspapers and glass bottles, none of these is made of nonrenewable materials that are in particularly short supply. The big savings from such efforts lie in reducing solid waste and the need for endless landfills and also, to some extent, in saving industrial energy use. What is crucial is that industry recycle rare materials, and this is beginning to happen because it happens not only to be good business but essential business. Substitutes can be found presumably for a great many materials as well, many of them one form or another of ceramics or plastics. And plastic comes from petrochemicals. Think of it this way: we have a 100-year supply of crude oil, and there is no telling how long the people of the world will need oil to make useful petrochemical *things*. So instead of husbanding such a valuable substance, we burn the vast majority of it to create motion (about 60 percent of oil used in the United States is for transportation) or momentary comfort in the form of localized heat or air-conditioning.

You don't even need to think cosmically (or even fret much about the greenhouse effect) to conclude that it is time to curtail humankind's use of fossil fuels.

But we must think about the greenhouse effect. Has it already started? How bad will it get? And what will it do to us? One thing is certain. The greenhouse effect will be a matter of climate but it will be felt biologically.

In the first place, the greenhouse effect has, happily, been present for eons. Without it, as we have seen, the Earth would be a huge iceball, and lifeless. Without the water vapor and carbon dioxide in the atmosphere that tend to reflect heat back to the Earth without preventing solar radiation from entering, the global temperature would be somewhere around –4 degrees Fahrenheit, or 36 degrees below the freezing point of water and 63 degrees below the present average global temperature over a year, which is 59 degrees. Not only that, supposing life could exist on a frozen planet, it would need to be a kind of life utterly alien to anything we now understand, since there would be no photosynthesis, thus no food chain, and not even any geologists or biologists to look into the matter. So we should pause and give thanks for the global greenhouse but then look seriously at what some people expect to be an unprecedented catastrophe.

It has been said many times: what we are going through now is a large-scale geophysical experiment, the likes of which the Earth has never experienced before. Within a couple of centuries, we will return to the atmosphere and the oceans all of the concentrated organic carbon that has been stored in sedimentary rocks for hundreds of millions of years as coal, natural gas, and oil. This sounds momentous, and it is. However, the annual emissions of carbon dioxide from these sources are actually quite small compared to the titanic amounts involved with the carbon cycle of the Earth, the natural exchanges in photosynthesis, respiration, and decay, and the natural exchanges between the atmosphere and the oceans. The point is that the natural processes are operating in a kind of equilibrium, or what might be called a balanced budget. There may be no way that these processes can accommodate the additional anthropogenic emissions.

Thanks to tiny bubbles of air trapped in the annual layers of polar ice, we have a record of carbon dioxide concentrations in the ambient air. The accompanying graph shows the increase in the past 250 years—a curve not at all unlike the curve of fossil fuel use in the same period. The second graph shows in greater detail the increase since 1958. The large annual oscillations reflect the grander global cycle: the uptake of carbon dioxide from the air by plants as they grow in the spring, and the return of carbon dioxide to the air as they decay. From such data, and given present and anticipated rates of energy use, it appears that the level of carbon dioxide in the atmosphere will have doubled by the middle or maybe the latter part of the next century.

The first suggestion that burning fossil fuels could have a climatic effect came from a Swedish scientist, Svante Arrhenius, who estimated toward the end of the nineteenth century that there would be an increase in global temperature of 3 to 11 degrees Fahrenheit for a doubling of the carbon dioxide level. Later estimates, in 1938 and 1956, put the temperature rise at 4 degrees and 7 degrees, respectively, in the event of doubling. Today we have estimates from far more sophisticated modeling of the atmosphere via computers. They range from 2 to 11 degrees, still a very broad range in prediction limits. Indeed, what may seem an embarrassingly broad range is the result of the fact that we now know a great deal more about the complexity of the climate—and one of the things we have learned is that we know very little about how it works.

Another thing we know is that there is more to all this than carbon dioxide. Certainly carbon dioxide is the major greenhouse gas that we are adding to the equation, not only in terms of quantity but overall potential effect. But there are others, notably methane and our old friends the CFCs.

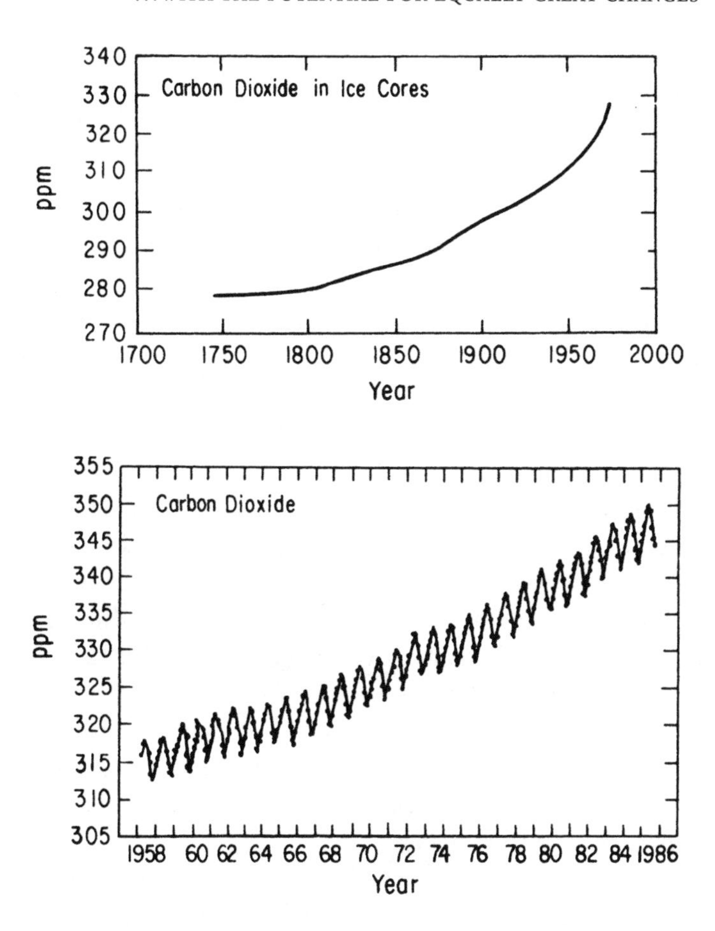

Above, the concentration of carbon dioxide in air trapped at different depths in polar ice. **Below,** the concentration of carbon dioxide in air as measured at the Mauna Loa Observatory, Hawaii, from 1958 to 1986. Units of concentration are parts per million (ppm). From Firor, 1990.

Since preindustrial times, carbon dioxide in the atmosphere has risen from 275 to 346 parts per million. In that same period, methane has risen from 0.75 to 1.65 ppm, and CFCs have gone from none whatsoever to about 650 parts per trillion. Methane arises chiefly from livestock feedlots and from the growing of rice. From the figures, methane and CFCs would seem like greenhouse peanuts—mere irritants. But this is not so. Methane is twenty times more potent as a trapper of heat than is carbon dioxide, molecule for molecule. And a molecule of CFC will trap *20,000* times the heat that a molecule of carbon dioxide does. In addition to those gases are nitrous oxides and ozone (which is a heat trap if found in the troposphere, the region below the stratosphere, which is about the only place where we want to find a lot of ozone).

When it comes to pointing the finger at various nations of the world as the big greenhouse villains, the United States heads the list, and it should. We use the vastly greater share of the world's fossil fuels, and not too efficiently. But when the emissions of all the greenhouse gases are factored in with a view to their potency as heat trappers, some surprises emerge. The United States, with its cattle as well as its power plants and automobiles, still comes in first, producing almost 18 percent of the problem. Next is what used to be the Soviet Union, producing about 12 percent, followed by Brazil. Thanks in large part to the burning down of forests (trees being major carbon sinks), Brazil adds more than 10 percent of the world's greenhouse gases. China and India—huge rice-producing regions—follow. If China, which sits on some of the world's largest coal reserves, decides to use them for a period of rapid industrialization, it could well become the major contributor to the greenhouse effect. Americans have much to feel bad about in all of this, but guilt is a commodity to be shared in various measures. As individuals, we are hoggish, but others are more so. In fact, on a per capita basis, the people who emit the most greenhouse gases are those who live in Qatar, United Arab Emirates, Bahrain, Canada, Luxembourg, Brazil, and Cote d'Ivoire, in that order, followed by the United States.

Meanwhile, to add to the confusion of climate forecasting, there are the emissions of sulfur dioxide, which, as we saw in chapter 2 on volcanism, tend to make a haze that inhibits incoming solar radiation and has a cooling effect on Earth. Some scientists have suggested that the effect from sulfur dioxide will nearly counterbalance the effects of greenhouse gases, which would seem to be a morally fragile achievement of harmony by pollution and waste. Then there is the problem of clouds. Low clouds can have a cooling effect by reflecting light back into space, but high clouds have a warming effect by trapping heat rising from below. In any global warming scenario, clouds are expected to increase in extent, and in some models the net cloud effect reduces the predicted warming from greenhouse gas doubling by anywhere from 4 to 10 degrees Fahrenheit. Beyond those matters lie the interchanges between atmosphere and oceans that are the ultimate sink for a great deal of atmospheric carbon dioxide. Few suggest today that enough is known about such dynamics to produce a completely reliable model.

It is not unnatural for geologically minded souls to look into the past for insight about the relationship between atmospheric carbon dioxide levels and global temperature. At present we have about 300 ppm. One hundred and twenty million years ago, carbon dioxide was six times more abundant in the atmosphere—1,800 ppm. And during the period that the carbon dioxide content of the atmosphere decreased to its present level, the temperature at low latitudes decreased from 82 degrees to 62 degrees Fahrenheit. From this geologic record, we can make an estimate for the temperature increase that would derive from a doubling of carbon dioxide (or its equivalent from all the culprits with their varying effectiveness). We had a temperature change 120 million years ago of 20 degrees with a decrease of 1,500 ppm, from 1,800 to 300 ppm. So $(T/20) = (300/1{,}500)$ or $T = 4$ degrees Fahrenheit, a figure that falls fortuitously but tellingly midway within most of the estimates from computer models.

Scientists no longer disagree on the question of whether the greenhouse effect has begun. Over the past century, there have been natural changes in climate that might have masked any anthropogenic events. In a major analysis of climatic effects from volcanoes (which tend to lower temperatures), El Niño, solar variation, and other natural causes were shown to not be enough, when all was added and subtracted, to account for the rise during the twentieth century of the average global temperature by 1.1 degrees Fahrenheit. That left only human activity to make the difference.

Within the vagaries of the various models used to predict the rise of average global temperature in the next century, we can reasonably expect, even plan on, an increase in global temperature of 4 to 8 degrees Fahrenheit in the middle to late part of the century. One such prediction, superimposed on the temperature record of the recent past, is shown in the accompanying graph, which also illustrates the important point in all greenhouse predictions: the extraordinarily rapid rate of increase in temperature compared to anything ever seen in the historical or geologic past.

The effects of such a global rise are as hard to predict as the size of the temperature rise and what its dimensions will be regionally. There is general agreement that the temperature will rise more near the poles and not very much at all near the equator, that midcontinental regions will become drier, and that the odds for violent storms, both inland and in the seas, will increase. One estimate suggests that the odds for extreme heat waves in the three cities modeled—Washington, Des Moines, and Dallas—would double or triple.

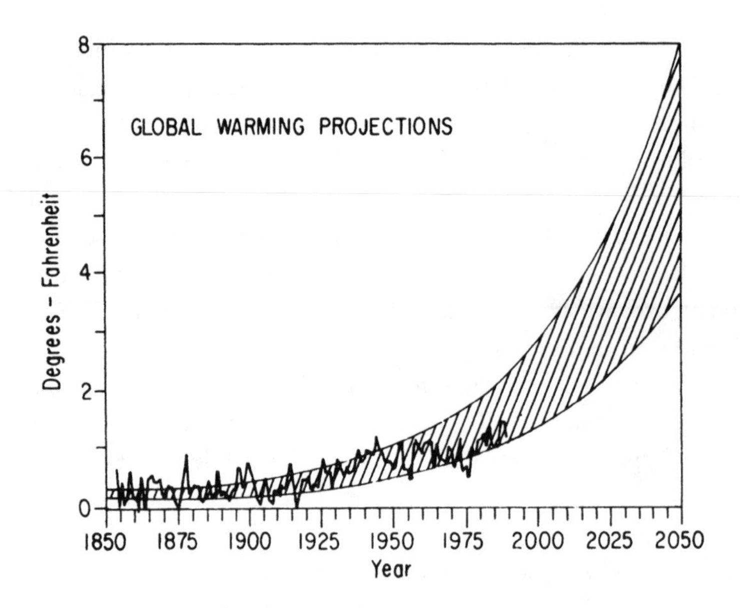

The range of global warming predicted by various computer models (diagonal region) superimposed on the historical record of temperature change. From Jones and Wigley, 1990.

Not all of the news is bad. Carbon dioxide is a necessary part of the photosynthesis process, and for certain crop plants like wheat, barley, rice, and potatoes a doubling of atmospheric carbon dioxide would "fertilize" them to the point that yields would increase by 30 percent or more. On the other hand, a doubling would have no particular effect on plants such as maize, millet, and sugar cane. More northerly lands would become more arable, while places like the American Midwest would become more arid, perhaps calling for the growth of different crop plants. A place like Iceland would have the climate of present-day Scotland and would be able to grow as much hay as Scotland does today. Many forecast a net gain for agriculture, though many adjustments will be necessary.

Today, most forecasters see the greatest agricultural dislocations and failures occurring in Africa and in some of the southern Asian nations, places where the population will grow at the greatest rate. Population scholars predict that by the year 2050, some nine billion people will inhabit the Earth, and almost all this growth will occur in the Southern Hemisphere. That is half again as many people as live on the Earth today. Many experts are optimistic that the Malthusian dicta will not apply, even if the global population reaches ten billion.

Others see a far gloomier outcome, especially with the predictable and unpredictable results of global climate change added to the mix. It seems fairly likely, for example, that many of the world's agricultural lands will experience not just drought but excessive rainfall and flooding, as has been happening with increasing frequency in the American breadbasket—the Midwest. The more climatologists and other scientists look into it, the more complexity they see in the climatic and ecological webs of which the biosphere consists.

Two decades or so ago, the head of NASA's Goddard Institute for Space Studies, James Hansen, appeared before the United States Senate to explain what was known and not known about the vast anthropogenic changes that had begun to occur and would continue to occur at an increasing rate unless the governments of the world took action. Among other events Hansen predicted, he pointed out that warming would be the greatest at the poles and that before too long, maybe three decades, the Arctic Ocean would be free of any ice in the summer months.

The loss of ice would mean that less sunlight was reflected back into the atmosphere and away from the earth's surface, creating what sounds nice—a positive feedback loop—but would be devastating. The less reflective ice and snow, the darker the sea and the land become, absorbing more heat, warming the waters and the land even faster, on and on, over and over.

In any event, twenty years later, in 2008, James Hansen spoke again to the Senate and in online commentary. He pointed out that what was occurring in the Arctic was just as he had predicted. The ice is diminishing and at a faster rate than had been predicted by even the most gloomy, removing vital habitat for the likes of polar bears, walruses, many seals, and narwhals, those single-"horned" creatures mistaken in earlier times as unicorns. (The horn is actually an elongated single tooth.) With land temperatures rising as well in the Arctic, whole villages are collapsing as the permafrost on which they rest thaws, leaving the land mushy. The same fate is occurring with roads, pipelines, and other important Arctic infrastructure. It was predicted that by the end of the Arctic summer of 2030 the north pole would be ice-free.

In another instance of positive feedback, the thawing permafrost opens up vast fields of peat that in turn emit methane into the atmosphere. Methane is not as prominent a greenhouse gas as carbon dioxide, but it has a more intense effect on the greenhouse.

The Arctic, Hansen said, is a tipping point, a situation where matters can and soon will be entirely beyond human remedy. If the nations of the earth (and particularly the United States, Russia, China, and India) continue on the same path as now, the world, Hansen warned, will suffer mass extinctions of plant and animal life to rival the mass extinction at the end of the Cretaceous Age, when the dinosaurs and great numbers of ocean species vanished forever from the earth. Entire ecosystems will collapse completely, and sea levels will rise dramatically and precipitously.

Currently, according to Hansen, the amount of carbon dioxide in the atmosphere is 386.7 parts per million, and it is rising at 2 parts per million annually. We are, Hansen believes, facing a time bomb with little or no time to defuse. Somehow, the people of the earth have to reduce greenhouse gases to their level in 1988, which was 305 parts per million. Otherwise we will experience a cascade of tipping points, of positive feedback loops: for example, the same reflective-dark problems could emerge in the Antarctic. If, he says, we go ahead with business as usual, much of that southern continent's ice shelves will plunge into the sea, and sea levels around the world will rise as much as six feet—a lot more than that if the Greenland glacier melts (which it is doing at a rate higher than accounted for in the 2007 report by the Intergovernmental Panel on Climate Change). Even the simple addition of warmer water to the oceans increases the rise in sea level, as warmer water expands. The Maldives, an island nation in the Indian Ocean, is rapidly disappearing under the waves. The world's marshlands—those purifiers of sea water and homes for an astounding number of species–will disappear along with most beaches. Large parts of many nations, including the United States, will likely be under water.

Hansen, who has emerged as a global climate change prophet whose prophecies have so far proved true, stirred up a great deal of controversy by suggesting that the leaders of the carbon-based energy industry should be jailed for crimes against nature and humanity—the crimes being the industry's creation and support of captive scientific laboratories that have worked so hard over the years to try to cast doubt on the entire topic of climate change, just as the tobacco industry did for years, stalling the inevitable. For, as noted, the petroleum age is coming to an end in the history of the Earth, what with what is called peak oil—the probability that we are now entering a production decline of the resource. One positive effect (not feedback loop) of climate change is that we may wind up changing our energy sources sufficiently broadly so that soon some petroleum may be left for other valuable uses, rather than simply being burned as fuel.

Hansen has always been out front on this huge and looming problem; although it has taken them a longer time, his colleagues in this matter, of which there are several thousand, are all in agreement on the obvious facts of the case. The only places where actual scientific disputes arise are over the details. And the details are more often than not found buried in elaborate computer models that attempt to explain the past as closely as possible and are then applied to the future based on the present. This leaves a lot of room for informed discussion.

Many in the print and electronic media operate within what they call the fairness doctrine, designed to maintain objectivity. Put simply, it calls for putting forth both sides of a story. In complying with their fairness doctrine when, over the past decades, the press needed someone to oppose what at least 98 percent of all climatologists agreed to about global climate change, the only place to go for an opposing statement was to those scientists who were on the payroll of the carbon-based energy industries. So the public was given the utterly false notion that there was great disagreement over even the existence of global climate change. It is high time that the major media learned to tell the difference between legitimate scientific research and the work of well-paid hacks.

Not that there is anything simple about all this. Even the quantities, the numbers involved in all of this are largely unimaginable for most people. For example, the Antarctic ice sheet is 14 million square kilometers, and it contains 30 million cubic kilometers of ice. How do you grasp that? How many hockey rinks could be made from all that ice? We authors have no idea, but it would come as no surprise if most of the continental United States would be a usable hockey rink from sea to shining sea. And with plenty left over for our hockey-loving brothers and sisters in Canada.

Another way to imagine the scope of changes that may well be in store is to look at a particular, even minor system in nature. There are many such examples. One oddball prediction is that feral pythons, pets set loose when they become, shall we say, a bit overbearing, are numerous in Florida (where everyone seems to go if they have an exotic pet that needs to be set free). It has now been set forth in serious scientific prose that feral pythons in Florida, with the warming of the American South, could well make their way as far north as Kentucky.

Of more concern (except to those race horse stable owners of Lexington, Kentucky, who one morning may find their money stud with a large python wrapped around its leg) is the likely fate of coral reefs around the world, which are already under devastating if somewhat inadvertent assault from overuse by tourists and by agricultural run-off. A question of management principles arises here, of the sort that will become ever more common. Once people know what they are doing to destroy a resource like coral reefs by allowing too many tourists to frequent them or allow the fertilizers they have applied to their crops to migrate and slosh around among the corals, poisoning them, who is to blame, and who is in charge? Surely the actions are no longer the inadvertent effects of innocent people. Today they know what they are doing. Knowledge is power, but it also is responsibility.

But let us look at an ecological association that depends on the great rhythmic cycles of the planet that have come into existence over millions of years of evolutionary trial and error. For this we introduce a relatively unprepossessing shorebird, known as the knot. Among shorebirds, like sandpipers and plovers dancing on the edge of the surf, terns arcing through the sky, and gulls in their endless intramural squabbles, the knot is not a stand-out. Even the endurance it demonstrates, migrating from southern South America to the Arctic and back each year, is hardly unique. But the knot's timing is immaculate—and utterly crucial to its existence.

One of the main stopping-off places for knots on their journey north each spring is the beaches of Delaware Bay, now mostly lined with what are called, with more than modesty, "cottages," mostly summer houses. But those who are there in early June wake up one morning to find the beach littered with what at first look like so many World War I helmets. Surely some kind of massacre has occurred. But no, these are all horseshoe crab females who have arrived from across the Atlantic to lay their eggs under the sands of the beaches. Many have become grounded high above the tide line and are doomed. Many more make it back into the waves and head off across the ocean, tailed by the smaller males who wait offshore while the females perpetuate their race, which has been largely unchanged since they came into being some 500 million years ago, the evolutionary spawn of sea scorpions and remotely spiders. What has this got to do with knots?

The horseshoe crabs are highly prolific when it comes to laying their eggs. They have to be, because the knots will arrive within a day or two at this stopping-off place, nature's version of a fast food joint, to probe the sand

and dine on horseshoe crab eggs, gorging themselves for the journey north to the Arctic. On their arrival there, they lay their eggs, which hatch just as the weather and other signals bring about a huge bloom of insects, easy pickings for baby knots. Thus does this entire association depend on fairly steady timing of climatic weather patterns and even the phases of the moon (the moon is in charge, as it were, of the horseshoe crab migration).

So what will the knots, now numbering in the millions, do when the Arctic insect blooms occur two weeks ahead of schedule, an abrupt and sudden wrenching of a complex tapestry of lives and timing?

They will face possible extinction. You do not need to be down to the last couple to go extinct.

In any event, the planet is full of such ecological arrangements, the products often of millions of years of evolutionary experimentation. Many of them will be severed as forests move north, but too slowly; as plains go totally dry; as species move up mountains, seeking the level where their particular temperature and ecosystem have moved—plants can move upward at about ten feet a year—until they reach the top and their ecosystem has vanished into the thin air above the crest.

Will the tiger survive? The greatest concentration of tigers is in a huge marshy area along the India–Bangladesh border called the Sunderbands. The tigers there spend most of their time in the water and are known to sneak up on a fisherman in his boat and take him away for dinner. With a sea rise of a few feet, the Sunderbands will be under salt water, and it is likely that the tigers will not be able to use it. Indeed, the denizens of most refuges that have been established around the world to save many of the most charismatic creatures will have to move on, but the refuges have to stay put. Every time one looks at the potential changes ahead from global climate change (along with the other pressures we put on the earth), it seems more complicated, more exhaustingly detailed, and, for some, discouraging.

But the more we can forecast, the better we will be able to cope—not just with wildlife but with what the American intelligence community has described as global climate change's effects on national security. They foresee huge migrations of people, refugees swarming into other nations' lands with the destabilizing of many political regimes, enormous humanitarian burdens to care for the hungry and dispossessed (more than a billion people fall into those two categories already). Stepping far beyond any scientific expertise, one wonders if democracies—usually a bit messy and often slow, thank heaven—will be able to deal with such upheavals, or whether dictatorships will be on the rise, desperate to make the trains run on time (to borrow an old fascist concern from pre-World War II Italy).

Some people are confident that solutions can be found to most of the problems ahead, and of course solutions of one sort or another will have to be found. Even fewer believe that by some spectacular engineering, we can put off, even conquer global warming. Huge mirrors in the stratosphere. Why not, by way of a joke, get the artist Christo, who is famous for covering large pieces of the landscape with cloth, to head for the Arctic and use bright white canvas. One logical suggestion has come from a man who earned the Nobel Prize for Chemistry thanks to his work on the chemistry of the ozone layer.

Paul Crutzen of Holland has suggested the injection of huge amounts—tons upon tons—of sulfate particles into the stratosphere which would presumably cool the earth in the same manner as volcanic eruptions. After all, when the Philippine volcano Pinatub erupted on June 15, 1991, it produced the largest volcanic cloud in the twentieth century, causing weather changes that lasted for years, including cooler summers and warmer winters. So global were its effects that it has been implicated in the death of corals in the Red Sea and, far less to be lamented, a snowstorm in Jerusalem.

But there is always risk, especially in so grand an experiment. Indeed, such huge engineering efforts, called geoengineering, are lodestars for unintended consequences. One such consequence, suggests the National Center for Atmospheric Research in Boulder, Colorado, is that the sulfate particles might also have the effect of slowing down the self-healing of the ozone layer, expected to be complete by about 2070. Indeed, the scientists from Colorado suspect that Crutzen's scheme would result in the destruction of three-quarters of the ozone over the Arctic.

On the other hand, Crutzen is among those who are looking at this Earth from the standpoint of human intervention and who have taken a leaf from the age-old work of geologists and paleontologists. The job is to break up the billions of years of history of the planet into distinct periods. The periods all have strange names that are hard for anyone but geologists and paleontologists to remember, like Ordovican and Cretaceous. Up to now they have told us that we and our ancestors going back some twelve thousand post-glacial years live and have lived in a period called the Holocene (which means "recent whole," whatever that means). Crutzen and his colleagues, taking note of the increasingly all-encompassing intervention of mankind in the affairs of the planet, including global warming, consider mankind's

Average Annual Temperatures for Some Cities along the East Coast of the United States and Their Temperatures after a 6°F Anticipated Increase by the Mid- to Late-Twenty-First Century

City	Temperature Today (°F)	Temperature, 2050–2100 (°F)
Portland, Maine	45	51
Boston	52	58
New York	54	60
Philadelphia	54	60
Washington, D.C.	54	60
Richmond, Virginia	58	64
Atlanta, Georgia	61	67
Jacksonville, Florida	68	74
Miami, Florida	76	82

effects the equivalent of a geological force. They suggest that we call the age we now live in the Anthropocene—the age of the human-induced whole. They suggest that it began, for the sake of argument, in 1784, when James Watt invented the steam engine and lakes in Great Britain and elsewhere began to show sudden changes in the chemical composition of the water. And, a bit scarily, they say that the Anthropocene will last some 50,000 years at least, that being their estimate of when we might get rid of the last excess of greenhouse gases.

So here we are in the Anthropocene, in charge. Of course we can't do very much about volcanic eruptions, earthquakes, and other such tremendous insults as described herein, except to put ever more effort into being ready to cope after they occur. But many of the other matters we have discussed in this book are well within the reach of human ingenuity and political will power. It is now up to us. We have a lot of work to do to make a new accommodation to this one planet where human civilization will thrive—or not. The time to start, as people who plant trees often say, was ten years ago.

9

The Most Fundamental Question Facing Mankind Today Is Whether Man Can Evolve to Live in Harmony with Nature

There is a popular bumper sticker that says: "If the people lead, the leaders will follow." This is as good a motto as any as we face what seem to be overwhelming global troubles. But humanity has operated out of optimism (if sometimes misplaced) through its history, and a realistic optimism is certainly something needed now.

To that end, in this book we have sought to put some of our current problems in a context that is geologic, noting that the Earth has been around a long time and will continue to exist for a long time to come. It has continually undergone changes—small and vast—on time scales from a year or less to several million years. In terms of geologic time, there have been "abrupt" extinctions, such as the Ice Age megafauna and the dinosaurs—great extinction events that took place amid the constant background buzz of evolution and extinction. We know something of the environmental stresses that caused these extinctions but are far from understanding them in detail. They are of scientific interest in themselves, but a better understanding of them would surely cast light on the anthropogenic environmental changes we now face.

There are natural changes that humanity has come to accept and live with. Those who live on a floodplain or near an active earthquake zone or a volcano, or build a house on an eroding seashore, do so knowingly and at their own risk. These changes are on a short time scale, as are such disasters as the flooding of the Indus delta region in Bangladesh from typhoons in the Indian Ocean, or the alternating periods of drought and plenty in the Sahel. There the tolls in human lives are measured in the hundreds of thousands—unhappy examples of Malthus's "vice and misery"—and in the present circumstances of overpopulation, there seems little to do by way of corrective action.

Far greater have been even earlier changes on the Earth: the eruption of Tambora that led to a global cooling, New England's year without a summer, and Europe's last great subsistence crisis; the faith-shattering earthquake of 1755 off Lisbon; and the eruption of Santorini, which wiped out Minoan civilization and gave Plato the idea for Atlantis. And before mankind's arrival, these kinds of events were even more powerful.

But human beings and, in particular, the last couple of centuries of their existence, have brought about a new type of environmental stress. The most outstanding characteristic of this stress is the rapidity with which it has grown. Virtually nothing in the geologic record can compare with these rapid changes: we are changing the Earth's environment far faster than natural forces have done in the past.

Arguably the first serious environmental hazard introduced by humanity was the London smog, for which we have now substituted photochemical haze. In what amounts to a century, our municipal, agricultural, and industrial wastes have polluted many of our major coastal and inland bodies of water. Such effects are in themselves local or regional, but they occur in virtually every region: they are global. Similarly, when a species becomes extinct it is a local event, but extinction of species is now a global affair, and is taking place at a rate that is greater than that at the end of the Cretaceous period. And, in the years to come there will be an observable increase in global temperature—between 4 and 8 degrees Fahrenheit—that will take place with a rapidity unheralded in geologic history. Mankind (as a species) will adjust, but much of the flora and fauna of the world probably will not.

Still, there are reasons for optimism. For example, after the Atlantic gray whale was extinguished, the California gray whale was put on the endangered list. It became a cautionary roadblock to continued exploitation not just of the whale but of its habitat. Oil companies were frustrated to apoplexy by not being allowed to drill offshore in certain places that are on the whale's migratory route. Now the California gray whale population has recovered to the point where the question is whether it should be taken off the list of endangered species. From a standpoint of policy, we are hagridden in this case by success—and ignorance.

A population of mammals like gray whales that was reduced to hundreds and now numbers in thousands may look healthy, but there is many a slip in such a numerical recovery. Many or all could be the offspring of a few sires, which means that they might well be severely inbred and actually still severely endangered. They could be much less fertile, and they could be suffering a great deal of infant mortality, the common results of inbreeding that stalk a species early on, before freak anatomy shows up.

Ignorance is the essence of the problem that humanity faces as it looks forward to the harvest it has sown in a mere eye-blink of Earth time, or what we call geologic time. The important thing to understand is that the Earth doesn't care a whit how we go about our affairs. It has whirled around the Sun for billions of years and will continue to do so regardless of our efforts. We need to understand the stresses human activity has caused to the Earth's ecosystems in order to understand how to take the proper corrective actions. This might seem to be an obvious truism, but it needs stating and restating because there is not currently the kind of sustained, directive environmental research effort that is needed. Environmental research has not, at the national level, been perceived as vital until recently; it has not been on the front burner of research priorities. Perhaps this is because environmental damage sometimes takes a long time—years or decades—to reach a critical state, so shorter-term research projects may take precedence. The environment has no elected representatives; no senator works in its behalf, as a Connecticut senator works for the continued health of the state's nuclear submarine shipyards, or as a California senator labors to assure that the state's aerospace laboratories are funded. There is also a catch-22 when it comes to environmental research. It hardly ever gets done because, when a severe environmental problem emerges, we need the answers *now*. For policy makers to suggest that more research is necessary before decisions can be made is seen as a cop-out. We can't wait for tomorrow. So when tomorrow's problems arrive, we don't know what to do about them. And, anyway, tomorrow's problems haven't arrived yet, and we aren't sure what forms they might take, so we don't need to do any research on them at this time.

Of course, examples do exist of environmental research that has been good and useful, as well as plenty of bad examples. There have been a number of Chesapeake Bay programs, each successive one bringing in new players who have to spend a certain amount of time reinventing the wheel. And there is the NAPAP fiasco—a $500 million data-collecting exercise that produced no science and no understanding. Science is not the accumulation of facts; it is the reduction of ignorance. Ignorance lies all around us when it comes to addressing the problems of the environment. The difficulties of reducing all that ignorance were aptly discussed by Juan Roderer, a geophysicist at the University of Alaska, in a 1990 essay in *Eos*, the weekly magazine of the American Geophysical Union, which we quote here at length.

> How can we sustain a public sense of the common danger of global change while remaining honest in view of the realities of scientific uncertainty? How can we nurture this sense of common danger without making statements based on half-baked ideas, statistically unreliable results, or oversimplified models? How can we strike a balance between the need to overstate a case to attract the attention of the media and the obligation to adhere strictly to the ethos of science?
>
> The task of achieving a scientific understanding of the inner workings of the terrestrial environment is one of the most difficult and ambitious endeavors of mankind. It is full of traps, temptations and deceptions for the participating scientists. We are dealing with a horrendously complex, strongly interactive, highly non-linear system. Lessons learned from disciplines such as plasma physics and solid state physics, which have been dealing with complex non-linear systems for decades, are not very encouraging. The first thing one learns is that there are intrinsic, physical limits to the quantitative predictability of a complex system that have nothing to do with the particular techniques employed to model it....
>
> Because of these serious limitations on our ability to study global change, we are faced with the dilemma of balancing the realities of scientific uncertainty with our social responsibility. Our problem is compounded by the fact that scientific results are gobbled up, often out of context, by policy makers and politicians as they come out, often untested or unverified, in piecemeal fashion. Most of these individuals are not trained as scientists; scientific results are translated into "understandable" terms by intermediaries who often bias the translation to suit their own political agendas. This can lead to billion-dollar solutions for million-dollar problems, and possible economic disaster, or to

> million-dollar solutions for billion-dollar problems, and possible environmental disaster.
>
> Do we need proof beyond reasonable doubt before important environmental policy decisions are made? By the time we have proof it may be too late. On the other extreme, should costly policy decisions be made on the basis of insufficient scientific information just as preventive measures? Decisions on how much scientific information is "enough" should not be left to politicians alone—or to scientists alone. Both groups must work together. Not only must science issues be brought to policy makers in clearly understandable terms, policy issues must be brought to the science makers in understandable terms!
>
> The situation is not unique to any single nation. It presents both a great dilemma and a great challenge to the scientific community worldwide. It gives an added dimension of awesome social responsibility to the need for international scientific cooperation.

As clearly as Roderer's essay states the problem, there is one major element missing—you, the public. There is a commonly invoked axiom these days: Think globally, act locally. It is good and necessary advice. The world needs every bit of environmental economy it can get—the recycling of cans, newspapers, and plastic containers; the insulation of houses and hot-water heaters; the use of aerated faucets, radial tires, nontoxic cleaning solutions, and high-mileage automobiles. Without such benign activity, there will be no solution to either local or global environmental problems, but even such benign activity on the part of every American household would not be enough. We, the people, have to think globally and act both nationally and globally as well.

To begin with, policy makers may be in a uniquely good position to identify problems that need resolution. But the American public has been far ahead of its policy makers on most environmental issues for a long time. Policy makers are public servants, and it is the public that is the board of trustees.

Scientists are in a uniquely good position to find out arcane things that are necessary to know to reduce ignorance and facilitate good policies. And how scientists go about doing their business is complicated, hard for some lay persons to understand. Science produces morally neutral knowledge.

Policy makers use that knowledge to make ethical decisions based upon it. So where does the public fit into the scientific enterprise?

In the ethics of ignorance.

Scientific knowledge is neutral knowledge, but there is an ethics to ignorance. A vast proportion of American science is funded by the United States government—some $130 billion a year, in fact. That seems like a lot of money, but it is by no means enough to fund all of the scientific quests that need funding. Currently there is a great deal of internecine warfare among scientists and policy makers as to how that not-large-enough pie is to be sliced up and meted out. Setting such priorities is an ethical matter; it puts a dollar value on the various ignorances that scientific endeavor could reduce.

Many scientists see the major villain as what is called big science, huge projects defined typically as costing $200 million or more. One estimate is that one-fourth of the entire federal research budget of the 1990s will go to create and operate some thirty big science projects, with the rest of the funds spread as thin as paper among the other tens of thousands of research projects proposed, many of which will never get off the ground at all.

What are some of these megaprojects? One was the superconducting supercollider, a huge atomic particle accelerator some fifty miles in diameter buried in the scrubland of Texas. It was designed to keep the United States in the forefront of particle physics. Originally estimated as a $4.4-billion project, by 1990 the projected cost had jumped to $8 billion. It was killed by Congress.

NASA, the most grandiose planner of mega-big science projects in the history of the human race, has asked the U.S. government for $37 billion from the federal science budget over the next decade to build a space station. Not much science would take place on the space station (in the sense that science is a matter of testing hypotheses). It is merely a costly engineering project that is one step toward the multitrillion-dollar goal of putting an American on Mars in some thirty years.

Astronomy is another participant in big science. To look into questions to do with black holes, to search for dark matter in the universe (some of which is theoretically missing), and to explore other questions such as the age of the universe will cost upwards of $4 billion. And, as noted, there is the suggestion that a new array of telescopes be developed to do sentry duty in the event of a meteorite attack.

There is nothing at all wrong with any such quest for knowledge, nothing wrong even with wanting the thrill of putting a human on Mars. Such quests have been called "mankind at its finest." It is a bit worrisome that most big science projects do not go through the gauntlet of the peer review system. But the main problem is that everybody involved knows that for the foreseeable future the United States budget for scientific research is not going to grow. In fact, taking inflation into account, it has not grown since the end of the Eisenhower administration. And since it is the public's money, the public has every right to play a role in how the money is spent. What ignorance is it most important to dispel? Here are some things we do not know:

- The fundamental nature of matter
- What it is in our nature and origins that make human beings so nasty to one another
- Where the farthest galaxy is
- The exact series of events in the life and death of stars
- How many species of plants and animals there are on the planet and how they interact in an ecosystem
- Which species we need and which we can get along without
- When and how the greenhouse effect is going to change the world
- How big the ozone hole might really get, and what that might do to the ocean and therefore the climate, or what to do about any of this
- What more than 95 percent of the oceans' floor looks like, how it came into being, and what processes go on at these abyssal depths
- How we will feed and otherwise sustain ten billion people on Earth when we can't handle the present six billion
- If there is intelligent (or even unintelligent) life elsewhere besides Earth
- How the oceans affect climate
- How birds migrate
- How to function without using up all the petroleum in the Earth
- How to function without any petroleum or all the other materials that may well vanish one day
- What makes a sociopath
- How the universe began
- Where to put radioactive wastes
- How to irrigate land without ruining it in a few centuries or less

The list could go on almost indefinitely, and in much finer detail. This book has pointed out a number of other ignorances not on this list. Any citizen, however scientifically uninvolved, will have missed here an entire category of ignorance—medical research. There are plenty of others. But any citizen could rank these and other areas of ignorance as to urgency—indeed, as to moral necessity—and hand the policy makers a list of priorities to guide the scientific community. At the top of the list should be what we might call the life-sustaining sciences: all those inquiries that need to be made to maintain life in its greatest possible diversity and at its highest possible quality. Anything else can wait. Why shouldn't physics and chemistry be made handmaidens to the science of Earth and of life? The public can in fact demand this, and such use of human brainpower would truly be "mankind at its best."

It is an illusion to imagine that the Earth is going to return to some pristine condition, even a condition that existed before the Industrial Revolution. Lake Erie, for example, will always be something of an artifact, as are most of the world's rivers. Our wildlife refuges are really outdoor zoos, and we had best learn to think of them as such, so that they can be properly managed. There is no real wilderness. We have opted for technology and it is not likely that we will return to a pretechnological world, nor should we. The ideal vision one finds in such magazines as *Mother Earth News* is all well and good, and much of the Earth's travails would be over if such visions came true. But there simply isn't enough livable space for everyone to live a life of back-to-the-land subsistence. Most human lives henceforth will be urban and they will be sustainable only if we develop the technologies (as well as the science) that will let them flourish. And this means new materials, more efficient engines and motors, industries that are more in the nature of closed systems (that is, reusing materials and releasing little to the environment), and—however gradually—a shift to renewable energy resources, the ultimate one being the Sun.

Why not, for example, invest as a nation in the further development of solar energy systems to break down water into hydrogen and oxygen, then use the hydrogen as fuel (with water vapor the main effluent) for autos, or to run fuel cells to power the other engines of civilization? And once that technology is at a usable level, why not give it to China in the hope that they won't find it necessary to burn their vast coal reserves? Why not tie our aid to needy countries precisely to the humane population control plan of educating women, empowering them to act locally?

It is time to think globally, and to act globally. And it is we, the people, who will have to insist that environmental problems be solved—for they are matters of public health. Governments and industry will follow. The end of the world is not upon us, far from it. Since Nagasaki, no nuclear device has been detonated in anger. The cold war appears to be over, and the nations of the world are rapidly opting for democratic forms of government, that is, governments that must be responsive to peoples' needs.

The time has come to recognize that the most pressing need is to learn to live in harmony with the planet and its resources, not simply to plunder and overrun it. It is not a time for panic, but it is time for widespread and steady, long-term concern. There are no quick solutions. It must be accepted that some regions will never be brought back to whatever pristine conditions once prevailed. But we can see to it that the list of such places is not irrevocably increased. The Earth will exist for a long time, but we are the ones who will have to commit ourselves to the long-term process of making it a habitable place for our grandchildren and their grandchildren.

We created the problem: we can solve it.

References

1. The Earth Is Still Hot and Mobile

Tambora: The Year without a Summer

Stommel, H., and E. Stommel. 1983. *Volcano weather*. Newport: Seven Seas Press. 177 pp.

Post, J. D. 1977. *The last great subsistence crisis in the western world*. Baltimore: Johns Hopkins University Press. 240 pp.

Laki: Acid Rain in Europe

Uyeda, S. 1971. *The new view of the earth*. San Francisco: W. H. Freeman and Company. 217 pp.

Vink, G. E., W. J. Morgan, and P. R. Vogt. 1985. The earth's hot spots. *Scientific American* 252 (4):50–57.

Sigurdsson, H. 1982. Volcanic pollution and climate: The 1783 Laki eruption. *Eos* 63:601–602.

Krakatoa: Colorful Sunsets

Francis, P., and S. Self. 1983. The eruption of Krakatau. *Scientific American* 249 (5):172–187.

Simkin, T., and R. S. Fiske. 1983. *Krakatau 1883*. Washington, D.C.: Smithsonian Institution Press. 464 pp.

Symons, G. J., ed. 1888. *The eruption of Krakatoa and subsequent phenomena: Report of the Krakatoa committee of the Royal Society*. London: Trubner and Co. 494 pp.

Meinel, A., and M. Meinel. 1983. *Sunsets, twilights, and evening skies*. Cambridge: Cambridge University Press. 163 pp.

Santorini: The Myth Maker

Heiken, G., and F. McCoy. 1984. Caldera development during the Minoan eruption, Cyclades, Greece. *Journal of Geophysical Research* 89: 8441–8462.
Ninkovich, D., and B. C. Heezen. 1965. Santorini tephra. In *Submarine geology and geophysics*, ed. W. F. Whittard and R. Bradshaw, pp. 413–452. London: Butterworths.
Watkins, N. D., R. S. J. Sparks, H. Sigurdsson, T. Huang, A. Federman, S. Carey, and D. Ninkovich. 1978. Volume and extent of the Minoan tephra from Santorini volcano: New evidence from deep sea sediment cores. *Nature* 271:122–126.
Kaloyeropoyloy, A., ed. 1971. Acta of the first international scientific congress on the volcano of Thera. Athens: Archaeological Services of Greece. 437 pp.
Luce, J. V. 1969. *The end of Atlantis*. London: Thames and Hudson. 224 pp.
Page, D. L. 1970. *The Santorini volcano and the desolation of Minoan Crete*. London: Society for the Promotion of Hellenic Studies. 45 pp.
Vitaliano, D. B. 1973. *Legends of the earth*. Bloomington: Indiana University Press. 305 pp.

Toba, Yellowstone, and Long Valley: Gigantic Eruptions

Ninkovich, D., R. S. J. Sparks, and M. T. Ledbetter. 1978, The exceptional magnitude of the Toba eruption, Sumatra: An example of the use of deep sea tephra layers as a geological tool. *Bulletin Volcanologique* 41: 286–298.
Smith, R. B., and R. L. Christiansen. 1980. Yellowstone Park as a window on the Earth's interior. *Scientific American* 242 (2):104–117.
Francis, P. 1983. Giant volcanic calderas. *Scientific American* 248 (6): 60–70.

2. . . . And from Time to Time Its Surface Moves Around

Haicheng and Tangshan: Chinese Earthquakes and Earthquake Predictions

Gutenberg, B., and C. F. Richter. 1949. *Seismicity of the earth*. Princeton: Princeton University Press. 273 pp.
Yong, Chen, Kam-ling Tsoi, Chen Feibi, Gao Zhenhuan, Zou Qijia, and Chen Zhangli. 1988. The great Tangshan earthquake of 1976. New York: Pergamon Press. 153 pp.
Rikitake, T. 1978. Biosystem behaviour as an earthquake precursor. *Tectonophysics* 51:1–20.
Rikitake, T., ed. 1981. *Current research in earthquake prediction*. Dordrecht: D. Reidel. 383 pp.

Eastern Sichuan, 2008

Gates, Alexander, and David Ritchie. 2007. *Encyclopedia of Earthquakes and Volcanoes*. New York: Facts-on-File.
Hough, Elizabeth. 2004. *Earthshaking Science*. Princeton, NJ: Princeton University Press.
Page, Jake, and Charles Officer. 2004. *The Big One*. Boston: Houghton Mifflin.
Online reference for Magnitude 7.9, Eastern Sichuan, China: www.earthquake/usgs.gov/eqcenter/eqinthenews/2008

Scientific background on the Indian Ocean earthquake and tsunami: www.iri.columbia.ed/~lareef/tsunami

San Francisco: Earthquakes along the San Andreas Fault and United States Earthquake Predictions

Lawson, A. C., ed. 1908. The California earthquake of April 18, 1906. Report of the State Earthquake Investigation Committee. Washington, D.C.: Carnegie Institution. 451 pp.

Gere, J. M., and H. C. Shah. 1984. *Terra non firma*. San Francisco: W. H. Freeman. 203 pp.

Wesson, R. L., and R. E. Wallace. 1985. Predicting the next great earthquake in California. *Scientific American* 252 (2):35–43.

Kerr, R. A. 1989. Loma Prieta quake unsettles geophysicists. *Science* 246: 1562–1563.

Olson, R. S. 1989. *The politics of earthquake prediction*. Princeton: Princeton University Press. 187 pp.

Sorel, E. 1990. Footnotes to history. *American Heritage* 41 (1):55–63.

New Madrid and Charleston: Largest Earthquakes in North America

Byerly, P. 1942. *Seismology*. New York: Prentice-Hall. 256 pp.

Fuller, M. L. 1912. The New Madrid earthquake. United States Geological Survey Bulletin, no. 494. 119 pp.

Penick, J. 1976. *The New Madrid earthquakes of 1811–1812*. Columbia: University of Missouri Press. 181 pp.

Johnson, A. C. 1982. A major earthquake zone on the Mississippi. *Scientific American* 246:60–68.

McKeown, F. A., and L. C. Pakiser, eds. 1982. Investigations of the New Madrid, Missouri, earthquake region. United States Geological Survey professional paper, no. 1236. 201 pp.

Dutton, C. E. 1889. The Charleston earthquake. United States Geological Survey, ninth annual report, pp. 203–528.

Bollinger, G. A. 1972. Historical and recent seismic activity in South Carolina. *Seismological Society of America Bulletin* 62: 851–864.

Lisbon: All Saints' Day in Portugal

Davison, C. 1936. *Great earthquakes*. London: Thomas Murby. 286 pp.

Kendrick, T. D. 1956. *The Lisbon earthquake*. London: Metheun. 170 pp.

Hecht, A. 1977. François Marie Arouet de Voltaire. *Poem upon the Lisbon disaster*. Penmaen, Lincoln, 34 pp.

3. There Have Been Frequent Flooding and Sea-Level Change Events on Earth

Gilgamesh: The Flood

Rosenberg, D. 1988. *World mythology*. Lincolnwood: National Textbook. 520 pp.

Kenyon, F. 1941. *The Bible and archaeology*. New York: Harper and Brothers. 310 pp.

Scablands: Major Flooding Events

Bolt, B. A., W. L. Horn, G. A. Macdonald, and R. F. Scott. 1975. *Geological hazards*. New York: Springer-Verlag. 328 pp.
Ahrens, C. D. 1988. *Meteorology today*. St. Paul: West Publishing. 582 pp.
Bailey, J. F., J. L. Patterson, and J. L. H. Paulhus. 1975. Hurricane Agnes rainfall and floods, June–July 1972. United States Geological Survey professional paper, no. 924. 87 pp.
Butzer, K. W. 1976. *Early hydraulic civilization in Egypt*. Chicago: University of Chicago Press. 134 pp.
Bretz, J. H. 1969. The Lake Missoula floods and the Channeled Scab-lands. *Journal of Geology* 77:505–543.
Baker, V. R. 1978. The Spokane flood controversy and the Martian outflow channels. *Science* 202:1249–1256.
Allen, J. E., M. Burns, and S. C. Sargent. 1986. *Cataclysms on the Columbia*. Portland, Ore.: Timber Press. 211 pp.

New York and Miami: Relative Sea-Level Changes

Walcott, R. I. 1972. Late Quaternary vertical movements in Eastern North America: Quantitative evidence of glacio-isostatic rebound. *Reviews of Geophysics and Space Physics* 10:849–884.
Cathles, L. M. 1975. *The viscosity of the Earth's mantle*. Princeton: Princeton University Press. 386 pp.
Milliman, J. D., O. H. Pilkey, and D. A. Ross. 1972. Sediments of the continental margin off the eastern United States. *Geological Society of America Bulletin* 83:1315–1334.
Officer, C. B., and C. L. Drake. 1982. Epeirogenic plate movements. *Journal of Geology* 90:139–153.
Hamblin, D. J. 1987. Sleuthing the Garden of Eden. *Smithsonian* 18:127–135.

4. . . . And Occasional Visitors from Outer Space

Halley's Comet: Our Regular Visitor

Chapman, R. D., and J. C. Brandt. 1984. *The comet book*. Boston: Jones and Bartlett Publishers. 168 pp.
Etter, R., and S. Schneider. 1985. *Halley's comet, memories of 1910*. New York: Abbeville Press. 96 pp.
Lancaster-Brown, P. 1985. *Halley and his comet*. Poole, UK: Blandford Press. 186 pp.
Guillemin, A. 1875. Les comètes. Paris: Libraire Hachette. 470 pp.

Tunguska: Meteor Impacts

Heide, F. 1964. *Meteorites*. Chicago: University of Chicago Press. 144 pp.

Hoyt, W. G. 1987. *Coon mountain controversies*. Tucson: University of Arizona Press. 442 pp.

Grieve, R. A. F. 1990. Impact cratering on earth. *Scientific American* 262 (4):66–73.

5. The Earth's Climate Changes on a Variety of Time Scales

Little Ice Age: Advance and Retreat of the Glaciers

Grove, J. M. 1988. *The Little Ice Age*. London: Methuen. 497 pp.

Lamb, H. H. 1982. *Climate, history and the modern world*. London: Methuen. 387 pp.

Sahel: Alternate Drought and Plenty in the Sub-Sahara

Glantz, M. H. 1987. Drought in Africa. *Scientific American* 256 (6):34–40.

Lamb, H. H. 1988 *Weather, climate and human affairs*. London: Routledge. 364 pp.

Williams, M. A. J., and H. Faure, eds. 1980. *The Sahara and the Nile*. Rotterdam: A. A. Balkema. 607 pp.

Nicholson, S. E., and H. Flohn. 1980. African environmental and climatic changes and the general atmospheric circulation in Late Pleistocene and Holocene. *Climatic Change* 2:313–348.

Historical Temperature Changes: Ups and Downs, Whys and Wherefores

Ahrens, C. D. 1988. *Meteorology today*. St. Paul: West Publishing. 582 pp.

Lamb, H. H. 1982. *Climate, history and the modern world*. London: Methuen. 387 pp.

Michaels, P. J. 1990. The greenhouse effect and global change: Review and reappraisal. *International Journal of Environmental Studies* 36:55–71.

Gleick, J. 1987. *Chaos*. New York: Viking. 352 pp.

Welander, P. 1967. On the oscillatory instability of a differentially heated fluid loop. *Journal of Fluid Mechanics* 29:17–30.

Officer, C. B., and C. L. Drake. 1985. Epeirogeny on a short geologic time scale. *Tectonics* 7:603–612.

Ice Ages: Milankovitch Cycles or Chaos Theory

Plummer, C. C., and D. McGeary. 1988. *Physical geology*. Dubuque: Wm. C. Brown Publishers. 535 pp.

Drake, C. L., J. Imbrie, J. A. Knauss, and K. K. Turekian. 1978. *Oceanography*. New York: Holt, Rinehart and Winston. 447 pp.

Imbrie, J., and K. P. Imbrie. 1979. *Ice Ages: Solving the mystery*. Cambridge, Mass.: Harvard University Press. 224 pp.

Geikie, J. 1895. The great ice age. New York: D. Appleton. 850 pp.

Hallam, A. 1989. *Great geological controversies*. Oxford: Oxford University Press. 244 pp.

Posmentier, E. S. 1990. Periodic, quasiperiodic and chaotic behaviour in a nonlinear toy model. *Annals Geophysicae* 8:781–790.

Geologic Climatic Changes: Temperature, Oxygen, and Carbon Dioxide

Berger, W. H., and J. C. Crowell, eds. 1982. *Climate in Earth history*. Washington, D.C.: National Academy of Sciences. 198 pp.

Berner, R. A., and D. E. Canfield. 1989. A new model for atmospheric oxygen over Phanerozoic time. *American Journal of Science* 289:333–361.

Sundquist, E. T., and W. S. Broecker, eds. 1985. *The carbon cycle and atmospheric carbon dioxide: Natural variations Archean to present*. Washington, D.C.: American Geophysical Union. 627 pp.

Berner, R. A. 1990. Atmospheric carbon dioxide levels over Phanerozoic time. *Science* 249:1382–1386.

6. . . . And on Rare Occasions There Are Changes in Its Community of Living Things

Pandemics: Bubonic Plague

Gottfried, R. S. 1983. *The black death*. New York: The Free Press. 203 pp.

McEvedy, C. 1988. The bubonic plague. *Scientific American* 258(2):118–123.

Mee, C. L. 1990. Medieval Europe and the killer fleas. *Smithsonian* 20(11):66–78.

Tuchman, B. W. 1978. *A distant mirror*. New York: Ballantine Books. 677 pp.

Ice Ages: Demise of Woolly Mammoth and Cohorts

Martin, P. S., and R. G. Klein, eds. 1984. *Quaternary extinctions*. Tucson: University of Arizona Press. 892 pp.

Greene, J. C. 1959. *The death of Adam: Evolution and its impact on Western thought*. Ames: Iowa State University Press. 388 pp.

Sutcliffe, A. J. 1985. *On the track of ice age mammals*. Cambridge, Mass.: Harvard University Press. 224 pp.

Stewart, J. M. 1977. Frozen mammoths from Siberia bring the Ice Ages to vivid life. *Smithsonian* 8(9):60–69.

Vereshchagin, N. K. 1974. The mammoth “cemeteries” of Northeast Siberia. *Polar Record* 17:3–12.

Cretaceous/Tertiary: The Great Dinosaur Extinction Controversy

Alvarez, L. W., W. Alvarez, F. Asaro, and H. V. Michel. 1980. Extraterrestrial cause for the Cretaceous/Tertiary extinctions. *Science* 208:1095–1108.

Officer, C. B., A. Hallam, C. L. Drake, and J. D. Devine. 1987. Late Cretaceous and paroxysmal Cretaceous/Tertiary extinctions. *Nature* 326:143–149.

Officer, C. B. 1990. Extinctions, iridium and shocked minerals associated with the Cretaceous/Tertiary transition. *Journal of Geological Education* 38:402–425.

Hallam, A. 1987. End Cretaceous mass extinction event: Argument for terrestrial causation. *Science* 238:1237–1242.

Keller, G. 1989. Extended period of extinctions across the Cretaceous/Tertiary boundary in planktonic foraminifera of continental shelf sections: Implications for impact and volcanism theories. *Geological Society of America Bulletin* 101:1408–1419.

Zinsmeister, W. S., R. M. Feldmann, M. O. Woodburne, and D. H. Elliot. 1989. Latest Cretaceous/earliest Tertiary transition on Seymour Island, Antarctica. *Journal of Paleontology* 63:731–738.

Alvarez, W., and F. Asaro. 1990. What caused the mass extinction: An extraterrestrial impact. *Scientific American* 263:76–84.

Courtillot, V. 1990. What caused the mass extinctions: A volcanic eruption. *Scientific American* 263:85–92.

Crocket, J. H., C. B. Officer, F. C. Wezel, and G. Johnson. 1988. Distribution of noble elements across the Cretaceous/Tertiary boundary at Gubbio, Italy: Iridium variation as a constraint on the duration and nature of Cretaceous/Tertiary boundary events. *Geology* 16:77–80.

Carter, N. L., C. B. Officer, and C. L. Drake. 1989. Dynamic deformation of quartz and feldspar: clues to causes of some natural crises. *Tectonophysics* 171: 373–391.

Other Geologic Extinctions: Similarities and Differences among the Geologic Extinction Events

Hallam, A. 1989. The case for sea level change as a dominant causal factor in mass extinction of marine invertebrates. *Philosophical Transactions of the Royal Society London* B325:437–455.

Huffman, A. R. 1990. An endogenous mechanism for extinctions. *Geotimes* 35: 16–17.

Loper, D. E., K. McCartney, and G. Buzyna. 1988. A model of correlated episodicity in magnetic field reversals, climate and mass extinctions. *Journal of Geology* 96:1–15.

de Jager, C. 1990. Science, fringe science and pseudo-science. *Quarterly Journal of the Royal Astronomical Society* 31:31–45.

7. Then Along Came Us and We Have Effected Vast Environmental Changes on a Local and Regional Scale

London and Los Angeles: Smog

Wagner, R. H. 1971. *Environment and man.* New York: W.W. Norton. 491 pp.

Wise, W. 1968 *Killer smog.* New York: Rand McNally. 181 pp.

World Health Organization. 1961. *Air pollution.* New York: Columbia University Press. 442 pp.

Stern, A. C., R. W. Boubel, D. B. Turner, and D. L. Fox. 1984. *Fundamentals of air pollution.* New York: Academic Press. 530 pp.

Lake Erie and Chesapeake Bay: Marine Pollution

Officer, C. B., R. B. Biggs, J. L. Taft, L. E. Cronin, M. A. Tyler, and W. R. Boynton. 1984. Chesapeake Bay anoxia: Origin, development and significance. *Science* 223:22–27.

Neilson, B. J. and L. E. Neilson, eds. 1981. *Estuaries and nutrients.* Clifton, N.J.: Humana Press. 643 pp.

Officer, C. B., and J. H. Ryther. 1980. The possible importance of silicon in marine eutrophication. *Marine Ecology* 3:83–91.

Wagner, R. H. 1971. *Environment and man*. New York: W.W. Norton. 491 pp.

Burns, N. M. 1985. Erie: *The lake that survived*. Totowa, N.J.: Rowman and Allanheld. 320 pp.

Minamata: Toxic Wastes

Smith, W. E., and A. M. Smith. 1975. *Minamata*. New York: Holt, Rinehart and Winston. 192 pp.

Officer, C. B., and J. H. Ryther. 1981. Swordfish and mercury. *Oceanus* 24(1): 34–41.

Hartung, R., and B. D. Dinman, eds. 1972. *Environmental mercury contamination*. Ann Arbor: Ann Arbor Science Publishers. 349 pp.

President's Science Advisory Committee. 1973. *Chemicals and health*. Washington, D.C.: National Science Foundation.

United States Senate, Committee on Commerce, Subcommittee on Energy, Natural Resources and the Environment. 1970. Effects of mercury on man and the environment. Hearings of May 8, pp. 1–92.

United States versus Anderson Seafoods Inc. 1978. Federal Register, 447 Federal Supplement, pp. 1151–1160.

Lipton, D. W. 1986. The resurgence of the U.S. swordfish market. *Marine Fisheries Review* 48:24–27.

Chernobyl: Nuclear Wastes

Häfele, W. 1990. Energy from nuclear power. *Scientific American* 263(3):138–144.

Krauskopf, K. B. 1988. *Radioactive waste disposal and geology*. London: Chapman and Hall. 145 pp.

Medvedev, G. 1991. *The truth about Chernobyl*. New York: Basic Books. 274 pp.

8. . . . With the Potential for Equally Great Changes on a Global Scale

Ozone Layer Depletion: Natural or Anthropogenic?

Firor, J. 1990. *The changing atmosphere*. New Haven, Conn.: Yale University Press. 145 pp.

Stolarski, R. S. 1988. The Antarctic ozone hole. *Scientific American* 258(1):30–36.

Acid Rain: Acid Lakes and Forest Devastation

Firor, J. 1990. *The changing atmosphere*. New Haven, Conn.: Yale University Press. 145 pp.

Fulkerson, W., R. R. Judkins, and M. J. Sanghvi. 1990. Energy from fossil fuels. *Scientific American* 263(3):128–135.

Mohnen, V. A. 1988. The challenge of acid rain. *Scientific American* 259(2):30–38.

Beamish, R. J., W. L. Lockhart, J. C. Van Loon, and H. H. Harvey. 1975. Long term acidification of a lake and resulting effects on fishes. *Ambio* 4:98–102.

Nriagu, J. O., ed., 1978. *Sulfur in the environment.* Part 2, *Ecological consequences.* New York: John Wiley and Sons. 482 pp.
Roberts, L. 1991. Learning from an acid rain program. *Science* 251:1302–1305.

Population Growth: The Major Unaddressed Problem

Bogue, D. J. 1984. Population. *Encyclopedia Americana* 22:402–408.
Malthus, T. R. 1798. *An essay on the principle of population as it affects the future improvement of society.* London: J. Johnson. 396 pp.
Winch, D. 1987. *Malthus.* Oxford: Oxford University Press. 117 pp.
Cloud, P., M. Bates, J. D. Chapman, S. B. Hendricks, M. K. Hubbert, N. Keyfitz, T. S. Lovering, and W. E. Ricker, 1974. *Resources and man.* San Francisco: W. H. Freeman. 259 pp.
Meadows, D. H., D. L. Meadows, J. Randers, and W. W. Behrens. 1972. *The limits to growth.* New York: Universe Books. 205 pp.

Natural Resources Depletion: Fossil Fuels and Minerals

Starr, C. 1971. Energy and power. *Scientific American* 225(3):37–49.
Masters, C. D., D. H. Root, and E. D. Attansi. 1990. World oil and gas resources—future production realities. *Annual Reviews of Energy* 15:23–51.
Fulkerson, W., R. R. Judkins, and M. K. Sanghvi. 1990. Energy from fossil fuels. *Scientific American* 263(3):128–135.
Cloud, P., M. Bates, J. D. Chapman, S. B. Hendricks, M. K. Hubbert, N. Keyfitz, T. S. Lovering, and W. E. Ricker. 1974. *Resources and man.* San Francisco: W. H. Freeman. 259 pp.

Carbon Dioxide Emissions: Global Warming or Not?

Ahrens, C. D. 1988. *Meteorology today.* St. Paul: West Publishing. 582 pp.
Firor, J. 1990. *The changing atmosphere.* New Haven, Conn.: Yale University Press. 145 pp.
White, R. M. 1990. The great climate debate. *Scientific American* 263(1):36–43.
Jones, P.D., and T. M. L. Wigley. 1990. Global warming trends. *Scientific American* 263(2):84–91.

Global Climate Change

Bolin, Bert. 2007. *A history of the science and politics of climate change.* New York: Cambridge University Press.
Dow, Kirstin, and Thomas Downing. 2007. *The atlas of climate change.* Berkeley: University of California Press.
Gore, Al. 2007. *An inconvenient truth.* New York: Rodale.
International Panel on Climate Change publishes multi-volume reports that are available from Amazon.com. They include a series of four titled *Climate Change* 2007 with subtitles as follows: *Synthesis Report*; *Physical Science Basis*; *Impacts, Adaptation, and Vulnerability*; and *Mitigating Climate Change.*

9. The Most Fundamental Question Facing Humankind Today Is Whether Humans Can Evolve to Live in Harmony with Nature

Environmental Changes and Time: Geologic and Human Perspectives (See previous eight chapters.)

Environmental Research: A Catch-22

Officer, C. B., L. E. Cronin, R. B. Biggs, and J. H. Ryther. 1981. A perspective on estuarine and coastal research funding. *Environmental Science and Technology* 15:1282–1285.

Roderer, J. G. 1990. The challenge of global change. *Eos* 71:1085.

Local and Global Anthropogenic Changes: Where the Problems Are

The Earth Works Group. 1989. *50 simple things you can do to save the Earth*. Berkeley: Earthworks Press. 96 pp.

Epilogue: Where Do We Go from Here?

Gore, A. 1992. *Earth in the balance*. Boston: Houghton Mifflin. 407 pp.

Index

ASHLAND COMMUNITY & TECHNICAL COLLEGE
3 3631 1132143 5